Appriopriate Technology Manifesto

Shifting The Forces of Production to Empower People and Protect The Planet

Appriopriate Technology Manifesto

Shifting The Forces of Production to Empower People and Protect The Planet

Edited by

John Trimble and Mammo Muchie

AFRICA WORLD PRESS
TRENTON | LONDON | CAPE TOWN | NAIROBI | ADDIS ABABA | ASMARA | IBADAN | NEW DELHI

AFRICA WORLD PRESS
541 West Ingham Avenue | Suite B
Trenton, New Jersey 08638

Cover design & Book design: LiteBook Prepress Services

Cataloging-in-Publication Data may be obtained from the Library of Congress.

ISBNs: 978-156902-840-7 (HB)
978-156902-841-4 (PB)

Contents

INTRODUCTION

Why are we pursuing an Appropriate Technology Manifesto

John Trimble and Mammo Muchie

As part of the International Network on Appropriate Technology's (INAT) 2020 strategic plan, we made the decision to publish an appropriate technology manifesto. The 9th International Conference on Appropriate Technology (9th ICAT) hosted a workshop on 'Appropriate Technology declarations and manifestos'. The 9th ICAT was strongly supported and funded by the National Research Foundation (NRF) of South Africa. As a follow on to the conference, NRF provided some funding to support the publication of this manifesto. Major ongoing financial and administrative support for this manifesto comes from INAT. Any financial proceeds from this publication go to INAT to support ongoing appropriate technology projects.

The roots of this effort go back to the presentation of the Appropriate Technology Declaration at a National conference on Appropriate Technology held in Washington DC in 2010. This AT declaration is a cornerstone of much of the work of INAT and anchors INAT's approach to technology policy. The AT Declaration reads as follows:

> We declare that, the resources of our planet, Earth, must be used to develop the technologies needed to meet the basic needs of our total population, empower people and at the same time preserve resources for future generations.
>
> We declare that appropriate technologies must be prioritized to meet the basic needs of all humankind. Clean air

and water, safe food, affordable accessible and available healthcare, adequate shelter, renewable energy, relevant education, and access to information and communication are basic human rights that can be provided for all of humankind today by devoting the planet's natural, financial and human resources to appropriate technology.

We declare that appropriate technology development and deployment honors cultural sensitivity, maintains ecological balance and fosters economic sustainability.

We declare that priority must be given to providing technology focusing on empowering people directly. An empowered population can more effectively use their creativity and critical thinking to improve the state of the planet and society and improve their quality of life. An empowered population is most capable of exercising democracy in all local, national and global contexts.

We declare that in the process of implementing Appropriate Technology, that the most disadvantaged populations must be given priority. We must dedicate special attention to the technological needs of the 'Global South' – the most underdeveloped countries.

We declare that in the process of implementing Appropriate Technology, that special attention must be given to the least empowered sectors of all communities - women, children and oppressed minorities.

We declare that a primary goal of the 'Appropriate Technology movement' is to use science and technology research, development, deployment, implementation and policy to work toward worldwide egalitarianism. We envision – a social reality where all women, men and children of our planet have equal access to resources to meet their basic needs and a controlling voice in how resources are to be distributed, and technology is to be developed.

This declaration is on our website – www.appropriatetech.net. We encourage organizations to share and support this declaration. As an outgrowth of the AT declaration, this AT manifesto seeks to provide details on how our society has arrived at the current juncture of technology inequality. It then attempts to address various technology sectors with suggestions on how various stakeholder communities can contribute to promoting appropriate technology. The suggestions include both technology policy and practice.

The manifesto consist of three sections. The first section consist of three chapters that address the history of technology, current tensions in technology deployment and the different stakeholder groups considered in this study. The second section consist of five chapters. These chapters address different production sectors: manufacturing, agriculture, mining and services. A chapter is devoted to water and sanitation as a particularly significant service sector. The third and final section of the manifesto consist of three chapters. One chapter is devoted to presenting a model for the ethical foundation of appropriate technology. Chapter 10 presents a networked innovation system that promotes appropriate technology across Africa. The final chapter reflects on a number of the key points brought out in the manifesto and points the way forward.

This appropriate technology manifesto is a living document. It is the first attempt by INAT at a task that belongs to a much wider audience and a broad cast of stakeholders. We realise that it has various shortcomings due to our limited resources and time constraints. Suggestions for improvements, that will make future versions more robust and more effective, are welcome.

We would like to thank the participants of the 9th ICAT that contributed to the workshop on the AT manifesto as well as the range of INAT members that have contributed over the years to this process of revising our vision of appropriate technology

Regards, John Trimble,
Founding president of INAT

CHAPTER 1

How Technology has Impacted Economic and Cultural Developments and Division

John Trimble

Science is organized knowledge and wisdom is organized life – E. Kant[1]

Different approaches are taken to present a historical trajectory of global technology development. The author presents a personal view on the role technology played in family setting and workplace setting over the years. This is to serve as the basis for developing a model for citizen self-examination of their engagement with technology. The chapter centers on the role of paradigms in science and technology development in general and particularly science and technology revolutions. An examination of Kuhn's classic work on scientific revolutions based on paradigms is a starting point. Schumpeter's position on the role of innovation as the determining paradigm in technology advancement is presented and critiqued. Most current literature presents the fourth industrial revolution (4IR) as the dominant paradigm on how technology is impacting society today. This outlook is critiqued as euro-centric and challenged by examining the impact of technology through an indigenous and colonized people's lens. All four industrial revolutions are examined, and their impact discussed. A synthesis of this analysis of different paradigms leads to the formulation of the conquest mentality paradigm. The chapter ends with a challenge to confront the conquest mentality and address the decolonization of technology education and policy, as well as technology research, development, and deployment.

Author's Experience with Technology in Family and Community settings

We start by getting personal. This section examines two mental models. First is a snapshot of my view of technology's impact on my life in 1961 at the age of eleven. Second is a mental model of the impact computing had on my life from 1968 to the present. In both cases, I seek to link the technologies experienced to a broader global picture of technologies' impact.

During my early childhood, technology in our household was limited: a radio; black & white TV; an iron, refrigerator, stove, toaster and a coal-burning furnace. The most pervasive technology, outside the house, was transportation technology: the old car my father had, buses, the Elevated train (the L), and the railroad (my father was a Pullman porter). The automobile industry and railroad industry shaped much of technology's impact in the early and mid-20th century, driving the steel industry and oil industry that provided the components and fuels for the growing transportation industry.

Fossil fuels dominate the energy production of most of the world. For most of the 20th century, coal played a larger role than natural gas or petroleum products. As late as 2005, coal accounted for more energy consumption than natural gas. The service economy of 1960 looked a lot different than it does today. Newspapers were delivered to your door on a daily basis as was the mail. Milk trucks and ice trucks arrived several times during the week. However, the Coal truck made a major delivery to our house every Fall. I cannot forget that delivery!

The coal was dumped in the alley behind our house. I had to use a wheel barrow to haul load after load from the alley and dump it in the coal bin (a room in our basement). Every winter morning I had to tend to the furnace. This meant getting up at 6 am, transferring ashes from the bottom of the furnace to the top, adding coal and stoking the furnace to make sure the right amount of heat was supplied. This was an important task, given the harsh winters of Chicago. During particularly cold winters, we would run out of coal. I would travel 4 blocks with my 7 year old younger brother to the Coal Yard to buy one or two 25 pound bags of coal. We would haul this back in our little red wagon.

Coal accounted for almost all of the fuel based CO_2 emissions up to mid-20th century. In 1961 it accounted for 5 billion of the 9.6 billion tons of CO_2 emissions globally. While its percentage reduced, in 2019 coal remained the largest contributor, contributing 14.6 of the 36.4 billion tons of CO_2 [1]. This is what I remember about technology during my childhood. Fortunately, by my high school days the coal furnace was replaced by a gas furnace making my life a lot easier. Coal mining was the most significant mining operation for most of the world in the 19th and 20th century in terms of mined material, numbers of workers and contribution to carbon emissions.

During that same period (1959-1962) the most significant technology related event at school was the periodic air raid drill. We were ushered from our classrooms to the basement and instructed to stand silently in front of our lockers. We were told this was a necessary practice to prepare us for a possible nuclear attack. Even at 10 years old this made little sense to me as a defence against a nuclear attack. These drills were part of a much larger campaign to justify massive spending on nuclear weapons research and deployment. It all started with the Manhattan project – the largest, most costly research endeavour of its time. President Roosevelt initiated this secret project in December of 1942 in an effort for the USA to become the first country to develop an atomic bomb. This represented an international escalation in destruction research. Operating out of public sight at its peak the project employed up to 130,000 workers. By the end of World War II it had cost taxpayers over 2.2 billion dollars [2]. The fruition of this massive research project was the bombing of Hiroshima and Nagasaki. In August of 1945 a nuclear bomb was released above both of these cities killing between 129,000 and 226,000 citizens. This act of inhumanity brought a swift end to WWII.

The war's end did not end the efforts on this destruction research. This was only the beginning. The Cold War saw "The United States spent more than $9 trillion on nuclear weapons between 1940 and 1996" [3, p. 156]. Dismantlement of the Soviet Union brought an end to the cold war. Many thought this would mean the end of nuclear armaments. There was a brief reduction in military spending. However in 2009, just twenty days after being awarded the Nobel Peace Prize, Barack Obama, the world's newest Nobel laureate signed into law an act initiating a trillion-dollar expansion of the US nuclear arsenal [3, p. 157]. "Obama earned the ironic sobriquet 'President Drone' by launching almost ten times as many strikes – 563- as Bush and Cheney had [3, p. 189].

Author's Experience with Technology in Workplace settings

Computers have played a significant part in my working life, starting with my first computer course in 1968. It was Fortran programming using punched cards and a Control Data 6000 series computer. At graduate school, the Computer science department at Stanford exposed me to fundamentals in artificial intelligence (AI), timeshare mainframe computing and mini-computers running programming punched on paper tape. Working at HP between 1973 and 1980, I experienced the advances in programmable handheld calculators, mini-computers and printer technology using micro-processors. During this time Silicon Valley saw the birth of Apple computing and the expansion of IBM, Xerox and Intel. In the 1980s working for Xerox, Amdahl and Incomnet, I participated in the early development of networked computing, use of the internet, office computing, the advent of the mouse, spreadsheets and windowing in computing. Incomnet was engaged in outsourcing programming to India. As a Senior analyst working in LA, I oversaw the programming work of programmers in India making 10% of my salary. What followed over the years – advances in the internet, the Worldwide web, parallel computing, distributed computing, cloud computing and the internet of things (IoT).

Advances in computing power led to advances in AI – pattern recognition, robotics, machine learning, artificial neural networks and expert systems. These advances brought us robotic run manufacturing, advanced drone technology, self-driving vehicles and pervasive online shopping and entertainment. These developments came at a staggering pace, producing mega-monopolies as never imagined. In the 19th century the saying was 'the sun never sets on the British empire'. In the 21st century the reality is 'the sun never sets for the likes of Apple, Amazon, Alphabet and Alibaba'.

Developing and Sharing Mental Models

Stakeholders sharing mental models can lead to a better understanding of the current situation and its historical legacy, as well as assist in the understanding and selection of policy alternatives. Articulation of mental models does have its limitations. Chapter 3 of the World Development Report is on mental models. It identifies the biases and shortcomings of individual mental models, explores their importance in understanding situations and making decisions. "Mental models matter for dcvclopment because they affect decision making. Since a great deal of policy is based on changing people's decisions—to save and invest, to enrol children in school, to be active citizens, to be honest—understanding the role that mental models play in individual decision making opens up the possibility of new levers for policy, while at the same time highlighting potential problems" [4, p. 62]. Mental models provide an understanding of how something works in the real world and form expectations about the future. The process of sharing mental model requires developing a conceptual model that can be shared. The mental model is illicit knowledge that consist of interrelated memories, causal beliefs and conceptual relationships combining information and knowledge constructs [5]. An effective knowledge elicitation process is required to draw out an individual's mental model and construct the needed explicit shareable conceptualization.

A collaborative effort at sharing mental models and linking to a broader agenda is important. Shughart points out

> "In order to avoid the unintended consequences of failed, exogenous expert-induced policy change, development

> economists, behavioral economists and institutional economists should continue the work of trying to integrate the idea of shared mental models and endogenous social change into their models. While models may become more cumbersome and less parsimonious, the resulting improvements in policy prescriptions and the relevance of their models for lasting economic and social change on the ground would certainly outweigh the perceived costs of a loss of scientific precision." [6, p. 387]

Senge's work on learning organizations addresses the importance of mental models connecting with a systems approach and shared vision [7]. Senge's strategy combines mental models and system thinking with the concepts of personal mastery, shared vision and team learning to form his five discipline approach His application of these five disciplines to maintaining a learning organization has proved successful across industry and educational institutions [8] [9].

Isolated presentations of mental models, such as the two presented by the author, may appear anecdotal. Individuals can apply this mental model conceptualization process to gain a better understanding of the role they play in technology development and utilization. This conceptualization should attempt to place their experiences in a broader national and global context. A broader understanding of the trajectories of technology development are needed to assure the relevance of such conceptualization. This must accompany any effort to develop a citizen's approach to self-examination of the role of technology.

The Organized Development of Science and Technology

Science is the organized study of the both the material and immaterial aspects of the world. This requires the collection of data and information through a systemic process. This process organizes knowledge in a structured manner to create theories, rules, algorithms, and processes. The organization of this information and knowledge into categories results in the different branches of science. Technology is created by applying science to create devices.

These devices are used to perform tasks and solve problems. The technology used to solve problems can contribute to the further development of science. Given this essential relationship between science and technology, it is important to examine how science has advanced. Thomas Kuhn wrote The Structure of Scientific Revolutions in 1962. He followed this with revised editions in 1970 and 1996. The different editions have been translated into numerous languages. Kuhn's work is seen as groundbreaking and by some the most significant contribution to theory of scientific development [10] [11] [12].

Thomas Kuhn distinguishes normal science from scientific revolutions. "Normal science is defined as puzzle-solving and addresses three classes of problems-determination of significant fact, matching of facts with theory, and articulation of theory" [13, p. 34]. Kuhn focuses on the importance of paradigms that guide scientific work. Scientists, individually and as research teams follow a set of paradigms that are worldviews applied to the dynamics of the science and research, they are engaged in. Global development can take leaps with scientific revolutions. These scientific revolutions come about and are integrally linked to paradigm shifts. [13]. While a paradigm shift is a characteristic of every scientific revolution, all scientific knowledge is not cumulative.

> "Kuhn argues that different paradigms are incommensurable because they involve different scientific language, they do not acknowledge, address, or perceive the same observational data nor they have the same questions or resolve the same problems, neither they agree on what counts as an adequate, or even legitimate, explanation" [11, p. 51].

The decisions on what problems to investigate, how to collect information, and what collected information is relevant are all critical decisions shaped by the paradigm used by scientists.

Science is associated with good sense and reason and the scientific method is associated with experimentation and testing of hypotheses, providing and testing explanation that lead to establishing and extending causal links. Science also must be concerned with uncovering fallacies [14]. The ability and willingness to accept fallacies is linked to the scientist's philosophical and ideological world view. These are grounding aspects of the paradigm used by any individual or team of scientists.

Schumpeter's view of Innovation and Entrepreneurship

Schumpeter's view of how technology develops to advance society is based on the position that economic development is driven by innovation and entrepreneurship [15]. This perspective on economic development should be understood as a macro-level paradigm of technology development. In a firm, an innovation can result in a new product that can increase the market share of the firm because the new product is not only better than the firm's old product but is better than the products of the competition. This innovation may come from a totally new firm. An entrepreneur develops their first product that is better than the products of existing firms. Their new firm and new product seize market share from established firms with outdated products. This paradigm proved inspirational. It speaks to the constant improvement of the capitalist market system and offers an opportunity to everyone to engage in entrepreneurship where the most innovative and persistent individuals are guaranteed to succeed. This paradigm is extended on a national level to suggest that nations which are most innovative will provide the best technology, techniques, and end products to the public and will be rewarded with wealth in the form of profits on a national basis in a global context. This innovation process is often referred to as creative destruction where an old process is dismantled to make way for the new innovative process or, a new innovative device makes an older device obsolete. These cycles of innovation increase productivity and provide higher living standards for a broad population and in the process prolong capitalism. Many viewed this as evidence that through technological innovation capitalism can sustain itself indefinitely. Schumpeter expressed the belief that these innovation cycles would eventually lead to democratic socialism replacing capitalism [16]. The recognition of this creative destruction has resulted in the promotion of entrepreneurship as a solution to inequality, a source of individual development and an avenue for job creation.

A criticism of Schumpeter's analysis lies in the lack of clarity on how this innovation process is organized and controlled. Paul Sweezy is critical of Schumpeter's approach and speaks to the importance of identifying the class base of innovation [17]. He takes a Marxist, historical materialist perspective. This perspective identifies the owners (capitalists) and workers (proletariat) as distinct classes that have an antagonistic relationship. Owners use their ownership rights to control research and production. Innovations come from organized scientific investigations and technology research. This is primarily done by major corporate or government research organizations. The USA Department of Defense spends billions on technology research and development. Major tech companies, Microsoft, Google, IBM, Apple, Facebook, Alibaba, Tencent and Amazon spend billions on research. They are the source of most innovation. The funding for innovation is controlled by the owners of multinational and monopoly corporations. These powerful entities decide what innovations will be further developed and marketed. The notion that millions of independent entrepreneurial individuals are advancing the economy through their innovative creations is a myth. Innovations are team efforts usually funded by major corporations. Most corporations are privately owned, or public stock companies owned primarily by the wealthy. Therefore, the wealthy fund and control the innovative process of creative destruction.

Global Paradigms

A major global paradigm justifying and advancing technology development is the concept of the industrial revolution. Literature abounds affirming the importance of the steam engine in advancing industry and shifting the importance from agriculture to industry and manufacturing. The initial industrial revolution is associated with the rise in the textile industry in Britain in the mid-18th century. Mechanization of weaving was followed by machine tools, steel manufacturing and railroads and shipping based on steam power. Steam based production increased productivity. Steam based transport decreased time to transport goods and people. Faster and safer travel across oceans and vast land masses escalated the process of settler invasion.

This first industrial revolution has been followed by three more. Each of the four industrial revolutions is characterized by advances in technology that significantly altered the landscape of global production and distribution [18]. Every industrial revolution is considered a leap ahead of the previous period. Some see this as a global verification of Schumpeter's analysis of the role of innovation.

The second industrial revolution is associated with the period of 1870-1930 and characterized by electric power and the internal combustion engine. It brought the technologies of radio, telephone, television, electric lighting, home appliances, the automobile and airplane [19]. Manufacturing employed electric power and highway construction reshaped the landscape across the world. After WWII, breakthroughs in computing and information theory led to the 3rd Industrial Revolution. Advances in digital computing had a ripple effect impacting one industry after another. The storage of information, manipulation of information and transfer of information transformed the workplace in industries and agencies across the planet, but primarily in the advanced industrial countries.

The fourth industrial revolution (4IR) has become a concern of politicians, academics, and businesses. Technologies associated with the 4IR will have impacts on jobs, consumption, human development, and the environment. The 4IR was first popularized at the World Economic Forum (WEF) in Davos in 2016. The WEF founder and president, presented 4IR has having three fronts: the physical, digital and biological domains. Most discussions and academic papers focus on the digital domain highlighting the impact of artificial intelligence, ubiquitous computing, the internet of things (IoT), machine learning and cloud computing. Technologies associated with the 4IR also include biotechnologies, additive manufacturing, neuro-technologies, geoengineering, space technologies, and virtual and augmented realities.

The concept of Industry 4.0 is often linked and sometimes confused with 4IR. Its focus is on advanced robotics, additive manufacturing and the extensive use of computing and AI across industries. All these technologies are designed to increase productivity. This technology required funding. Therefore, it primarily benefited large companies and multinational corporations. The intent of increased productivity was to increase profits. Increased productivity is only profitable if the cost of the new technologies and techniques is less than what is gained by the increase in productivity. This requires the reduced cost of labour to offset the additional costs in technology. This can only come with fewer workers or lower wages. An alternative is to pass the profits on to the workers. Increased productivity works best with an expanding market. However, if profits are not passed on to the workers, fewer workers or lower paid workers means that the market shrinks. This was the fear expressed at Davos. Many participants in the WEF fear that a shrinking consumer base will force businesses to shut down and increase unemployment leading to a downward economic spiral, community unrest, the failure of capitalism and revolution. They contend that a controlled rollout of 4IR technologies with creation of new services to provide employment can instead allow capitalism to thrive.

One scenario of future developments of technology focuses on the centrality of artificial intelligence (AI) in the future of 4IR. In the future, AI developed across a range of technologies will have an increasingly significant impact on economic and social development. Lee suggests that the AI revolution takes place in four dimensions (or waves). They are internet AI, business AI, perception AI and autonomous AI. "Using internet AI, Alibaba (or Amazon) can recommend products you're more likely to buy, Google can target you with ads you're more likely to click on, and YouTube (or Netflix) can suggest videos that you're more likely to watch" [20, p. 107]. Business AI takes advantage of the massive data collected over decades by a range of businesses to employ sophisticated algorithms and data mining to uncover patterns and make decisions, outperforming practitioners and in many cases experts. Internet and business AI are considered well established. Perception AI has just began to develop, while autonomous AI is barely out of the gate. Perception AI is grounded in pattern recognition. Recognition of voices, faces, situations and structures has improved tremendously over the years, and expected future improvements will provide new unforeseen applications. Companies are just scratching the surface of autonomous AI. Truly autonomous robots and vehicles will transform transportation, mining and manufacturing, reducing the need for workers in a range of fields, while improving workplace safety, production speed, and product quality. Governments and large corporations are devoting significant resources to AI validating this perspective of AI as the driving factor in technology development. The most powerful governments and companies focus on applying AI to gain global dominance. Some believe this is likely to lead to increased global inequality with China and the USA leaving other countries behind [20].

A number of sources challenge the WEF narrative of the 4IR. There are reservations over if these technology advances even represent a new industrial revolution. They can be viewed as a continuation of the 3rd Industrial revolution. A broad set of challenges are presented by a community of sociologists highlighting that it a buzzword totally isolated from the social and economic realities of the people [21]. Barton goes even further and claims 4IR is based on discredited technological determinism. "The promise of 4IR is overblown and its perils are underappreciated. There are compelling reasons to reject—and even actively oppose—the 4IR narrative" [22, p. 31].

Technology Advancement Guided by Colonialism

The discovery of the Americas led to an escalation in colonial ventures crossing the Atlantic Ocean. This required developing the shipping industry along with associated manufacturing industries to make everything from rope and sails to cannons. The conquest brought Europe gold and silver. More importantly, the conquest brought a massive amount of land under the control of select European countries. First plantations on the Caribbean Islands provided the wealth to fund expansion in Europe. This required slave labor. The enslavement and transport of Africans across the Atlantic was the 2nd stage of this colonial conquest effort. The technology required for both the conquest of land and people was largely of a military nature. Williams speaks to this process in detail, providing clear evidence that capitalism was built on the slave trade, slavery, and the stealing of land [23].

> "if we want an honest assessment of the Industrial Revolution it is impossible to separate industry from colonial violence and genocide. ... Millions of Europeans flooded into North and South America, leaving space for population growth in Europe. If increased demand was a key driver of industrialization, there is no honest way to discuss this without including the expansion into the West. As more commodities were needed, science and industry had to transform to keep up with demand." [24, p. 33]

Europe's population in 1700 was 98 million. In 1900 its population was over 300 million. Space was less of a factor in population growth. Increased wealth brought to Europe allowed Europeans a better quality of life compared to other regions. This more likely aided in population growth. The wealthy in England operated plantations in Jamaica. The wealthy in France operated plantations in Haiti. The wealthy in Portugal and Spain operated plantations and mines in Central and South America. These very profitable slave operated plantations and mining operations brought untold wealth to Europe. This wealth created the drive for commodity development. Profits from the slave trade and slavery gave Europeans the money to build factories and money to buy commodities. "It was the capital accumulated from West Indian trade that financed James Watt and the steam engine" [23, p. 102].

The process of transforming science and industry in Europe could only come about with profits from conquest. Conquest could only take place with the superiority of military technology.

Competition among European countries for these massive profits from colonialism heightened their drive to expand military technology to control these colonial resources. Conquest of land and people was justified through racism, elitism, patriarchy, and nationalism. Cecil Rhodes exemplified this conquest mentality in Africa. He set out to link the British conquest from Cape Town to Cairo.

The Conquest Mentality as Ideological Principle Guiding Technology Advancement

Pre-conquest technology development focused on basic needs, providing food, safe water, clothing, shelter, and essential healthcare. Conquest mentality brought about the focus on conquest technology. When hunter civilizations turned on other civilizations this represented a backward turn for humankind. This motion coincided with the rise in patriarchy. Male hunters seized harvests, livestock, and women from rival groups. In some cases, this was an act of desperation to avoid hunger. In other cases, it was a conscious effort to secure a better life for their people at the expense of groups they felt no allegiance to. The rivalry spurred on the development of weapons, defensive structures and both offensive and defensive strategies to gain and maintain resources.

Conquest in the Americas is strongly associated with the conquistadors from Spain and Portugal who initiated the earliest conquests in the Caribbean, Central and South America and Mexico. What started as trade expeditions reverted to expanded efforts to extract precious metals. Military superiority was a major factor, but significant in many cases was the indigenous population falling victim to European diseases [25].[2] The Spanish and Portuguese were overwhelming men. In Mexico male settlers outnumbered female settlers 10 to 1. In Brazil the ratio was 100 to 1 [25, p. 17]. They were men on a mission seeking wealth. Settlers endured the hardship of transatlantic travel with one goal in mind, wealth. They came in numbers. It is estimated that between 200-250 thousand settlers came to the Indies in the sixteenth century from Spain alone. Many were peasants with farming skills, but with little interest in small scale farming. They had bigger dreams, which required aggressive conquest of the land to make quick fortunes [25, p. 7]. The land was ripped apart in search for gold and silver. Next came clearing vast land areas for lumber and space to set up plantations. This aggression on the land was the second form of conquest.

Conquest technology advanced with colonialism. The global colonial adventure needed technological advances in weapons, battleships, and fortresses to extend and maintain control of land and resources. First, these cross-continental colonial endeavors brought precious metals (gold and silver) and slaves to Europe. This provided the wealth that brought on the industrial revolution, in turn, creating the global inequality that persists today. This brought on the creation of race, racism, and white supremacy. A cyclical process developed. White supremacy and white privilege fed on itself to maintain and increase inequalities. Technologies of conquest, oppression and violence were further developed to suppress all effort for justice and equality. Advances of the industrial revolutions are glorified and projected as the center point of technology advancement concealing the truths of the extensive nature of technologies of oppression [26]. These technologies of oppression and violence were responsible for bringing to Europe the wealth required to bring about the first industrial revolution. Continued development and use of military technology assured the dominance of the British empire and the rise and dominance of the USA empire. Historically, it has been military domination that guaranteed the powerful nations and corporations their wealth and profits.

> "Military technology is usually the first to be borrowed by every society, because the penalty for failing to do so can be immediate and fatal; but the West seems to have been preternaturally receptive to new technological innovation, and the equally vital ability to respond to it, soon became an established feature of Western warfare" [27, p. 2].

European nations competition for colonial possessions started in the 16th century and continuing into the 20th Century. The combination of competition and shared innovation allowed these European nations to advance in military technology over Asia, Africa, and the Americas.

"The triangular trade made an enormous contribution to Britain's industrial development. The profits from this trade fertilized the entire productive system of the country" [23, p. 105][3]. While the profits from slavery and the slave trade advanced Europe technologically, it created instability in Africa retarding technological development. While militarily inferior, the warring forces in Africa did play a role. Their role was devastating for Africa as a whole contributing to the slave trade and divisions in Africa. These warring forces saw this capitulation as their only source of survival.

> "In 1788, West African warlords informed European observers 'that the sole object of their wars was to procure slaves, as they could not obtain European goods without slaves, and they could not get slaves without fighting for them'. Clearly peoples who fought to enslave rather than to exterminate their enemies would, like the indigenous inhabitants of the Americas, south-east Asia, and Siberia before them, prove ill prepared to withstand the unfamiliar tactics of destruction employed against them by the Europeans." [27, p. 6]

The control of transatlantic conquest required the expansion and further technological development of the Navy. Ships to transport slaves, precious minerals, manufactured goods, and people. However, most important were the warships to protect this cargo and control the sea routes. [4]

> "Furthermore, war – and especially that very commercially minded and middle-class organization, the British Navy – contributed even more directly to technological innovation and industrialization. Its demands were not negligible: the tonnage of the Navy multiplied from about 100,000 in 1685 to about 325,000 in 1760, and its demand for guns grew substantially," [28, p. 28]

The importance of naval superiority persisted into the 20th century. The tonnage of a steamship grew from 3,000 tones in the 1860's to 45,000 by 1914. The naval race required advances in steam engines, large scale metal structures and guns. Guns became bigger, faster, and more powerful. What started with the 10th Century Asian invention of gunpowder by the middle of the 20th century had advanced to nuclear armed mega-ships.[5]

> "From 1500 to 1930, European empires were synonymous with naval power. ... Britain's development of history's first four-ocean navy helped it defeat its Bourbon rivals, France and Spain in the global Seven Years' War (1756-63) and slowly strangle Napoleon's continental empire (1803-15), leading to its century of global dominion." P.161 [29, p. 161]

The ship building spree required construction of massive ship building facilities. This became the precursor to modern manufacturing industry.

The first mass production of manufactured metal goods was the weapon's industries. Mass manufacture of bullets required precision and speed that set the tone for future manufacturing of consumer goods.

> "By the seventeenth century, before iron had begun to be used on a great scale in any of the other industrial arts, Colbert had created arms factories in France, Gustavus Adolphus had done likewise in Sweden, and in Russia, as early as Peter the Great, there were as many as 683 workers in a single factory." [30, p. 89]

Wars have been central to the control of resources and the current global inequality. The control over the most advanced military technology has largely dictated the outcome of wars [31]

The shift to air warfare

The 20th century saw a shift from naval warfare to air wars. Fighter planes and bombers dominated WWII. They made possible the seizure of territories by ground forces. The dropping of atomic bombs on Japan at the end of WWII signaled a new day in the significance of air warfare.

> "the Vietnam conflict was the biggest air war in military history. By 1971, the United States had dropped 6.3 million tons of bombs on Indochina – double the total of 3.3 million tons the Allies dropped in all theaters during World War II. The air force and navy flew nearly two million combat missions, or sorties, in Indochina, more than the 1.7 million flown in World War II." P.162 [29, p. 162]

Bombs, missiles, and drones have dominated the warfare in destroying Libya and Iraq and waging 20 years of violence in Afghanistan. Military bombs and missiles keep Palestine in a constant state of disruption.

To maintain military superiority while minimizing their casualties, the USA has led the way in the development and deployment of robot-controlled vehicles and drones.[6] At the time of 9/11 in 2001, the USA government spent about $350 million on drones. By 2013, they spent well over $5 billion a year on drones [32, p. 4]. "By 2025, the United States will likely deploy advanced aerospace, cyberwarfare, and digital surveillance technology to envelop the earth in a robotic matrix theoretically capable of blinding entire armies or atomizing an individual insurgent" p. 182 [29, p. 182]. These efforts at military control represent the continuation of the conquest mentality into the 21st century.

Nations that emerged strongest were those that developed military technology and exercised this conquest mentality. The scientist, technician and soldier all played a role. Headrick sums up the relationship between technology and warfare. "In short the partnership between the soldier, the miner, the technician, and the scientist is an ancient one. To look upon the horrors of modern warfare as the accidental result of a fundamentally innocent and peaceful technical development is to forget the elementary facts of the machine's history." [30, p. 87]

Wealth generated from conquest ventures made possible the expansion of consumption. This demand drove investment in industry to produce consumer goods of all types. However, the demand for military products was always most profound. This was driven by the highly competitive nature of conquest and the fact that so much of warfare technology and products were destroyed in the process of conquest or often quickly rendered obsolete.

> "The most wanton and luxurious household cannot compete with a battlefield in rapid consumption. A thousand men mowed down by bullets are a demand more or less for a thousand more uniforms, a thousand more guns, a thousand more bayonets: and a thousand shells fired from cannon cannot be retrieved and used over again. In addition to all the mischances of battle, there is a much speedier destruction of stable equipment and supplies. Mechanized warfare, which contributed so much to every aspect of standardized mass-production, is in fact its great justification." [30, p. 93]

Conquest of the Land

The conquest mentality extends to how settlers approached the land. Not only did mining leave scars wherever settlers landed. Vast forests were decimated to export lumber to Europe. Ships that brought timber exports to Europe provided cheap transport for more settlers to travel to the Americas. In 1808 Napoleon's 'continental Blockade' prevented Britain from getting much needed timber from Europe. Britain shifted, timber imports from North America went from 6% in 1803-7 to 74% in 1819-23. Tonnage went from 21,000 tons in 1802 to 110,000 in 1815 and then 340,000 in 1819. "This meant that in the five years 1815-19, the number of potential passenger berths and cargo space to North America suddenly tripled." [33, p. 109]

The expansion of railroads provided labour for many settlers and migrants and attracted many newcomers. What is often overlooked is the destructive and disruptive pattern that came with rail expansion. In North America, the expansion of rail is linked to near extinction of the buffalo. Destruction of wildlife and environment was an integral part of the conquest by settler colonialism. It was not just a conquest of the indigenous people and the conquest, transport, and enslavement of Africans. It was the destruction of the land and its ecosystem in the name of taming the wilderness. "Buffalo hunting became a wildly popular sport for white people ... Men from all over the country boarded trains headed west in order to shoot buffalo with .50-caliber rifles from train windows. They killed thousands of buffalo a day, leaving the animals' lifeless bodies where they fell on the plains to rot." [34, p. 22]

Construction of railroads fuelled the steel industry and coal and iron mining. It allowed for the escalation of settler colonialism with faster, safer and cheaper transport of new arrivals. Rail technology was developed to extend this process of conquest. Toward the end of the 19th century, the top nations in rail miles were all English speaking settler colonies. "In 1875, the top five nations in terms of rail miles per capita were the United States (with 1,922 miles of rail per million people), New Zealand (1,350), Canada (1,159), Australia (998) and Britain (527)" [33, p. 108]. The railroad was an effective means of troop transport to quickly come to the aid of settlers in the conquest process. It was not the only infrastructure development associated with conquest. Construction of forts were significant building efforts central to the conquest process. They represented such a significant architectural and organizational structure in settlements that today one can find numerous cities and towns around America named after forts.[7] "The business of defence became complicated in proportion as the tactics of offence became more deadly: road-building, canal-building, pontoon-building, bridge-building became necessary adjuncts of warfare" [30, p. 88].

The extremes of Conquest

The efforts to conquer the smallest particles known were played out in extreme form by the Manhattan project, mentioned earlier in this chapter. It employed over 130,000 people, most of the leading physicists in the world, to what end? The unleashing of the most destructive weapon ever used. Today efforts continue to conquer even smaller particles. Currently, the Large Hadron Collider (LHC) is the world's largest atom smasher. It is located in France near the border with Switzerland and operated by the European Organization for Nuclear Research (known by its French acronym, CERN). There are plans for a new collider that would be four times bigger and 10 times more powerful than the current LHC. This effort would require multinational collaboration to build a $24-billion facility consisting of several labs and a circular tunnel with a circumference of 100 kilometres (62 miles). Is this appropriate spending when tens of millions of the world citizens are without proper water and sanitation?

At the other extreme is the billions spent on conquering space, our last frontier. The Soviet Union was the first country to put a man in spacc in 1961. The response from the USA was a massive increase of funding to the National Aeronautics and Space Administration (NASA). Between 1962 and 1970, NASA received the equivalent of 288 billion in 2020 constant dollars. The 2021 NASA budget of 23.3 billion is detailed on the Planetary Society website [35]. The worlds two wealthiest men, Elon Musk and Jeff Bezos are competing for the lead in the "last conquest". Bezos' Blue Origin spent more than $2 million lobbying in 2020 for a cut in billions of funding for projects with NASA. Currently on the line is a $10 billion project to return NASA astronauts to the moon. [36]. Billionaires Paul Allen and Britain's Richard Branson are also spending millions in the privatization of the space race. Earlier in 2021, Branson, Musk and Bezos all sent private spacecraft into space at the cost of millions [37]. How can anyone justify spending billions on this space competition?

The voice of the oppressed

It is important to understand the role of indigenous and colonized people in creating a paradigm shift to redefine technology development and redirect technology deployment. These are the citizens of the world who are most aware of the damage done by the conquest mentality. They are most aware of the polluted waters, toxic air and decimated forests that are the results of the technology misadventures around the world. Indigenous leaders and organizations have been at the forefront of calling for climate justice. Evo Morales, president of Bolivia at the time, pointed out that after the financial crisis of 2008

> " While the United States and the European Union allocate 4.1 trillion dollars to save the bankers from a financial crisis that they themselves have caused, programs on climate change get 313 times less, that is to say, only 13 billion dollars." [38, p. 80]

Many that have suffered when the USA and Europe have resorted to using their advance military technology are speaking out. There is a call to divert funds from military technology and adventures. In 2009, the president of Bolivia pointed out:

> "The budget of the united States is 687 billion dollars for defence. What do they want to designate for climate change in order to save life and humanity? 10 billion dollars. It's an embarrassment. According to information which we obtained the budget for the war in Iraq is 2.6 trillion. Trillions of dollars to go and kill Iraqis and Iranians in Iraq. What do they want to allocate to us for climate change? Ten billion dollars. Well, it isn't fair." [38, pp. 92-93]

It is not just the governments that focus on inappropriate technologies. Major corporations have a long history of irresponsibility in caring for the environment. Oil multinationals are at the top of the list of companies that have caused damage through their choice of techniques and technologies. "Fossil fuel industry continues to boost production, drilling, and fracking, developing one new megaproject after another, as it boasts ever louder about how much it cares" [39, p. 47]. When corporations pushed the governments of Canada and the USA to pursue expanding the gas pipeline from Canada through to Texas, it was the indigenous peoples of North America that organized. They were outraged by the potential damage they realized the pipeline could do to the water systems and the broader environment. People have called out corporations to be more accountable for environmental damage. They pushed governments to regulate corporations. In turn corporations funding politicians pushed the government to allow corporations to self-regulate. In self-regulating corporations pushed the onus onto the public. Various projects engaging the public in climate regulation, collecting recyclables, etc. held the activist citizen accountable to mitigate climate change and implement environmental stability. The corporations continued to do more damage [39]. The damage done through this conquest of nature through extractive ventures goes beyond the surface damage and the impact on climate change.

> "Climate change destroys animal habitats by causing wildfires, hurricanes, floods, thawing, and drying, bringing animals and humans into closer proximity. That in turn, increases the chances of humans contracting zoonotic infections. It was a forest fire connected to a drought, for example, that caused fruit bats in Malaysia to relocate to trees on farms, where they infected pigs with Nipah virus. The pigs then infected farmers, and more than a hundred people died." [39, p. 48]

The causal linkage of damages to the planet and its people caused by conquest mentality is extensive. Much of this damage is yet to be recognized due to the delay in the cause-and-effect dynamics. The abandonment of these inappropriate technologies and techniques requires the total rejection of the conquest mentality and agenda. We cannot continue to view the planet, its land, its waters, its plants, its animals, its people, its air, and its natural resources as things to be conquered, dominated, and exploited.

Conclusions

Examination of indigenous technology (and culture) is important in uplifting the most oppressed. However, this can only be the start. Empowered indigenous people and the most oppressed communities, to chart a future of sustainable technological empowerment, is needed to counter this conquest mentality. Utilizing human centered technology, which was much of the focus before colonial invasion, can help in addressing the injustice imposed by conquest. This technology redirection must focus on basic needs, not profit for the wealthy and corporate owners.

A call for appropriate technology is a call to empower people. To uplift the oppressed there must be broad engagement in the process of decolonization. Colonization was made possible through conquest. Conquest was only possible through the technologies and techniques of conquest. The machinery and machinations of conquest were and continue to be possible because of the ideological perspective of conquest – the conquest mentality.

Those in academia can start by decolonizing our educational institutions. A reexamination of how we teach our youth the history of human exploration and advancement must start with recognition of this flawed approach of conquest. The decolonization process is much more complex than most realize. As Andrews points out

> "Decolonizing the curriculum has to mean more than being a bit critical of Kant or adding some diversity to reading lists. If, and this is a big if, we want to build an anti-racist society we need to completely rethink the underlying basis of the knowledge that produces the world." [24, p. 18]

We must seriously rethink our approach to science, to technology, to engineering, to development itself. The revolutionary transformation of the curriculum is just a part of the necessary global, comprehensive decolonization process that is just starting.

Bibliography

[1] H. Ritchie and M. Roser, "Our World in Data," 2021. [Online]. Available: https://ourworldindata.org/emissions-by-fuel. [Accessed 10 September 2021].

[2] Office of Legacy Management, "Manhattan Project Background Information and Preservation Work," [Online]. Available:

https://www.energy.gov/lm/manhattan-project-background-information-and-preservation-work. [Accessed 29 July 2023].

[3] C. D. Conner, The Tragedy of American Science, Chicago: Haymarket Books, 2020.

[4] World_Bank, "World Development Report: 2015," World Bank, Washington DC, 2016.

[5] J. S. Holtrop, L. D. Scherer, D. D. Matlock, R. E. Glasgow and L. A. Green, "The Importance of Mental Models in Implementation Science," *Frontiers in Public Health*, vol. 9, no. July, p. doi: 10.3389/fpubh.2021.680316, 2021.

[6] W. F. Shughart II, D. W. Thomas and M. D. Thomas, "Institutional Change and the Importance of Understanding Shared Mental Models," Kyklos, vol. 73, no. 3, pp. 371-391, 2020.

[7] P. Senge, The Fifth Discipline, New York: Currency doubleday, 1990.

[8] P. Senge, A. Kleiner, C. Roberts, R. Ross and B. Smith, The Fifth Discipline Fieldbook Strategies and Tools for Building a Learning Organization, New York: Doubleday, 1994.

[9] P. Senge, N. Cambron-McCabe, T. Lucas, B. Smith, J. Dutton and A. Kleiner, Schools That Learn A Fifth Discipline Fieldbook for Educations Parents, and Everyone Who Cares About Education, New York: Doubleday, 2000.

[10 J. Kuukkanen, "Rereading Kuhn," International Studies in the Philosophy of Science, vol. 23, no. 2, pp. 217-224, 2009.

[11] T. Orman, "Paradigm" as a Central Concept in Thomas Kuhn's Thought," International Journal of Humanities and Social Science, vol. 6, no. 10, pp. 47-52, 2016.

[12] J. Mayoral, "Five Decades of Structure: A Retrospective View," Theoria, vol. 75, pp. 261-280, 2012.

[13] T. Kuhn, The Structure of Scientific Revolutions, 3rd ed., Chicago: The University of Chicago Press, 1996.

[14] S. S. Carey, A Beginner's Guide to Scientific Method, Belmont California: Wadsworth, 1994.

[15] J. Schumpeter, The Theory of Economic Development, Routledge Classics Edition ed., London: Routledge Classics, 2021.

[16] J. A. Schumpeter, Can Capitalism Survive?, Barakaldo Books, 2020.

[17] P. Sweezy, "Professor Schumpeter's Theory of Innovation," The Review of Economics and Statistics, Vols. Vol. 25, No. 1, no. No. 1 - Feb, pp. 93-96, 1943.
[18] K. Schwab, The Fourth Industrial Revolution, Geneva: World Economic Forum, 2016.
[19] K. Schwab, Shaping the Fourth Industrial Revolution, Geneva: World Economic Forum, 2018.
[20] K.-F. Lee, AI Super-Powers China, Silicon Valley, and the New World Order, New York: Houghton Mifflin Harcourt, 2018.
[21] T. Ngwane and M. Tshoaedi, The Fourth Industrial Revolution: A sociological critique, Johannesburg: Jacana Media, 2021.
[22] C. J. Barton, "The Fourth Industrial Revolution Will Not Bring the Future We Want," IEEE Technology and Society Magazine, pp. 31-33, September 2021.
[23] E. Williams, Capitalism and Slavery, New York: G. P. Putnam's Sons, 1980.
[24] K. Andrews, The New Age of Empire, How Racism & Colonialism Still Rule the World, New York: Bold Type Books, 2021.
[25] A. Hennessy, "The Nature of the Conquest and the Conquistadors," Proceedings of the British Academy, vol. 81, pp. 5-36, 1993.
[26] A. Alveren, The Economics of Military Spending: A Marxist Perspective, New York: Routledge, 2019.
[27] G. Parker, The Cambridge History of Warfare, Second Edition ed., Cambridge: Cambridge University Press, 2020.
[28] E. Hobsbawn, Industry and Empire The Birth of the Industrial Revolution, 2nd ed., New York: The New Press, 1999.
[29] A. W. McCoy, In the Shadows of the American Century: The Rise and Decline of US Global Power, London: Oneworld Publications, 2019.
[30] D. R. Headrick, The Tools of Empire: Technology and European Imperialism in the Nineteenth Century, Oxford: Oxford University Press, 1981.
[31] D. Vine, The United States of War A Global History of America's Endless Conflicts, from Columbus to the Islamic State, Oakland: University of Calfornia Press, 2020.
[32] W. M. Arkin, Unmanned Drones, Data, and The Illusion of Perfect Warfare, New York: Little, Brown and Company, 2015.

[33] J. Belich, Replenishing the Earth The Settler Revolution and the Rise of the Anglo-world, 1783-1939, Oxford: Oxford University Press, 2009.
[34] I. Oluo, Mediocre The Dangerous Legacy of White Male America, New York: Seal Press, 2020.
[35] The Planetary Society, "NASA's FYI 2021 Budget," 2021. [Online]. Available: https://www.planetary.org/space-policy/nasas-fy-2021-budget. [Accessed 10 October 2021].
[36] C. Davenport, "A rivalry that extends to the moon," Washington Post, 22 May 2020.
[37] C. Davenport, The Space Barons, New York: Hachette Book Group, 2018.
[38] E. M. Ayma, The Earth does not belong to us, we belong to the Earth, La Paz: Ministry of Foreign Affairs, 2010.
[39] J. Balkan, New Corporation How "Good" corporations are bad for democracy, New York: Vintage Books, 2020.
[40] C. Bishop, Ships, 2017 ed., London: Amber Books, 2017.
[41] S. Lindqvist, "Exterminate All The Brutes", New York: The New Press, 2007.
[42] F. Gosling, "The Manhattan Project: Making of the Atomic Bomb," United States Department of Energy, Oak Ridge, TN, 1999.

Endnotes

1 Kant is considered a leading philosopher of the 18th Century. He emphasized the intension of one's action, the importance of moral duty and that reason establishes this moral duty. Kant also advocated the inferiority of Africans and other people of color based on biology and climate. [24] points out numerous examples of his racist writings and actions. However, I open with this quote because the relationship between science, knowledge, technology, and wisdom is key to the future of humankind and the welfare of our planet. Organized life must consider all humankind and the sustainability of the ecosystem on Earth.

2 Europeans brought diseases that native Americans had no immunities. Mexico's population declined from 25 million in 1512 to 1.9 million by 1580, while Peru went from 10 million in 1530 to 1.5 million by 1590. [25]

3 The triangular trade sent ships with surplus manufactured goods to Africa, slaves from Africa to the Caribbean. Slave produced sugar back to England. Each stage of the triangle was profitable to England. The trade to Africa went from textiles to guns.

4 As far back as 17th century Britain and France had warships with over 100 large cannons, 200 feet long and displacement of over 1000 tons. The Santissima Trinidad the largest warship of the 18th century, built in 1769 with a displacement of 4572 tons and 130 guns [40].

5 "The Chinese had discovered gunpowder in the tenth century and had cast the first cannon in the middle of the thirteenth. But they felt so safe in their part of the world that, from the middle of the sixteenth century onward, they refrained from participating in the naval arms race.

Thus the backward and poorly resourced Europe of the sixteenth century acquired a monopoly on ocean-going ships with guns capable of spreading death and destruction across huge distances. Europeans became the gods of cannons that killed long before the weapons of their opponents could reach them.

Three hundred years later, those gods had conquered a third of the world. Ultimately, their realm rested on the power of their ships' guns" [41, p. 47].

6 "Under a super-secret "black" contract, Lockheed Martin built 20 RQ-170 Sentinel drones with 65-foot wingspan each cost $200 million. After Iran captured on in 2011, the Pentagon worked with Northrop to develop an even larger, more sophisticated drone, the RQ-180 with a wingspan of 130 feet. The project cost an estimated $3 billion with each drone costing $300 million" [29, pp. 184-86].

7 Ten largest cities named after their forts, city name and population: Fortaleza, Brazil = 2,571,896; Fort Worth, TX = 741,206; Fort Wayne, IN = 253,691; Fort Lauderdale, FL = 165,521; Fort Collins, CO = 143,986; Fort Smith, AR = 86,209; Fort-de-France, Martinique = 85,667; Fort Myers, FL = 65,725; Fort McMurray, AB = 61,374

Fort Portal, Uganda = 54,275 (https://panethos.wordpress.com/2015/11/15/most-populous-fort-cities-on-the-map/ last visited 10 October 2021)

CHAPTER 2

A System Thinking Paradigm to Technology Development

John Trimble

This chapter provides a system thinking perspective to technology development and deployment. Historical materialism is presented as the application of dialectical materialism to society, culture, production, and reproduction. The major dialectic in historical materialism is the tension between the forces of production and relations of production. Technology is presented as a key aspect of the forces of production. How this shapes class struggle and ideological consciousness, as well as relationships of production is examined. The changes in technology force changes in relations of production. Those in control of production guide technology development to maintain and advance their control over resources. Historical materialism and system thinking are used to advocate policies promoting technology that empowers communities and promotes ecological balance.

The dominance of the conquest mentality continues to shape technology policy. A systemic approach must be formulated to marshal forces to advocate appropriate technology. In formulating this approach, the biggest challenge is to overcome the dominance of conquest thinking. Historical materialism offers a materialist theory to examine the socio-economic impact of technology. It requires an examination of economic production and distribution, with regards to the role technology plays in inequality. System dynamics (SD) provides a methodology for quantifying the systemic relationships and positing scenarios that will examine real and potential consequences of continuing or rejecting the conquest mentality. SD uses computational numerical analysis to represent the differential equations that reflect the non-linear feedback relationships identified.

Dialectical Materialism, Science, and Technology

Chapter 1 examined several paradigms associated with the development of science and technology. A clear link is established between capitalist technology and the role of slavery and settler colonialism [1]. The chapter ended by laying bare the extent of the conquest mentality and its fundamental role in the development of the technological capitalism that dominates the world today. In this chapter we start with our focus on dialectical materialism as a world view.

Dialectical materialism is a scientific approach to examine different forms of matter and material existence. It can be explained based on three fundamental principles [2]. The law of contradiction (the unity and conflict of opposites) explores the tension between the primary components of material reality being examined. In material science, the study of the strengths of different metals involves this law. A court of judges, faced with making a ruling, is a social setting where the law of contradiction is applied. The second law is the transition of quality into quantity and of quantity into quality. This accounts for all qualities of matter having quantitative counterparts and all quantitative characteristics of matter leading to qualitative expression. A color representing a quality has a corresponding wavelength that is captured as a quantity. A certain quantity of watts applied to a light bulb produces a certain quality of brightness. Countries with a gross national income (GNI) per capita between $1,036 and $12,535 are classified as middle-income countries by the World Bank. The third law of dialectics is the negation of negation. The process of negation is the destruction of some aspect of a thing and the bringing to dominance of another aspect. Negation of negation is then destruction of the new aspect and bringing about an aspect that may resemble the earlier aspect or be a new aspect. This process of negation of negation is ongoing for all material reality. British advances in military technology led to its negation of the prior dominance of

Portugal and Spain as military powers. The 20th Century saw the USA surpass Britain in military power bringing about a negative of the earlier negation. The USA technological supremacy will see its negation, furthering the process of negation of negation. Hopefully, society will replace it with a more human focused technology agenda, that is a negation of the conquest mentality.

Applying dialectical materialism to technology development can be used to explain a range of science fundamentals. Dialectics explains the forces in tension and motion, internal and relative, that features in all of matter. This is useful in advancing understanding physics from the study of the universe to subatomic particles. Material science and chemistry are also advanced by applying the fundamentals of dialectical materialism. We find infinitely different material expressions to investigate. The investigations into these material expressions are orchestrated by the forces that control the resources to explore science, advance technology and develop and deploy new technologies. Science has advanced over the years to develop a wide range of tools, techniques, and machines to further explore: 1) the forces of tension that make up different forms of matters; 2) the impact that external forces of different types have on specific types of matter; and 3) the characteristics of the different types of motion internal to specific instances of matter. Decisions are made to develop and apply these tools to explore the fundamentals of matter and to transform these fundamentals into useful technologies and end products. These decisions are not random.

The decision to investigate the particulars in science and to apply science to develop technologies is a conscious choice. This choice is not made on an individual basis. Communities of scientists, technicians and engineers make organized decisions leading to advancement of science and technology. The decision-making process, that drives technology development and deployment, is determined by those individuals and institutions in control of the resources to enact this complex process. Multi-national corporations (MNCs) spend billions on research and development (R&D). Governments allocate billions to R&D. A significant portion of this government funding is used for R&D and deployment of military technology.[1] This technology is used to maintain existing conquests or pursue new conquests.

Historical Materialism Fundamentals

Historical materialism is the application of dialectical materialism to the social, economic, and political arena of history and everyday life. It is a paradigm for our collective human interaction in production, reproduction, and cultural development.

> "This force, historical materialism holds, is the method of procuring the means of life necessary for human existence, the mode of production of material values – food, clothing, footwear, houses, fuel, instruments of production, etc. – which are indispensable for the life and development of society." [3, pp. 27-28]

Central to this application is the identification of the plenum of forces in tension at any given historical epoch. We seek to find a common characterization across these historical epochs. The most common characterization focuses on the tension between forces of production and relations of production.

The forces of production are the people engaged in production, their skills and the tools and machines they use in the production process. Relations of production are 1) how the people relate to each other in the production process, 2) how people are impacted by the means of production, and 3) how people relate to the ends of production. Ends of production are the products and services provided and profits from the production process. The means of production include the tools and machinery used for production (which are part of the forces of production), the raw materials put into the production process and the space used for production [4]. The people engaged in production are divided. There are those who own the means of production and those who work the machines for a daily wage. This is the main class division which has existed throughout history. Today, the owners are the capitalists of society. They want to increase their profits from production and increase their control over the production process. This is distinctly opposed to the interest of the workers employed in production. The workers also want more control over production and more of the profits. Capitalists and workers are clearly in an antagonistic relationship. Each will support technology decisions that further their class interest and vehemently oppose technology that is against their interest.

We are accustomed to dividing production into services and goods. Also, production is divided into consumer goods and services, and goods and services for production itself. The manufacture of pots, pans or shoes are goods for consumers. Consumers get dental, medical, and educational services. In contrast, auditing a company is a service to a business and an aid to production. A letter sorting machine is manufactured to directly assist the post office (production) not the final consumer. The platinum mined in South Africa and Zimbabwe goes as a raw material to other production processes, not the final consumer. Another (less discussed) division of work is that between productive and unproductive labor. Productive labor contributes directly to creating or enhancing a material product or service. Unproductive labor is used to assure cooperation and maintain the current system. Soldiers, security guards, and bank clerks checking credit ratings are examples of unproductive labor. It is essential to examine technology's role in all three dimensions. Technology plays a role in the forces of production, defining skills, tools, and machines of production. Technology is also a key aspect of the final products of production. Products of production to consumers are more and more of a technological nature. Products of production that are used in production also are largely of a technological nature.

How does the technology used in production impact the well-being of workers and the stability of our planet? It is known that technologies from hydraulic fracturing for oil and gas, to offshore drilling have negative impact on land and sea. Packaging technology using plastics has resulted in billions of tons of plastic waste on land, in rivers, lakes and the ocean. Communities that host these flawed technologies suffer. The workers who use and produce these flawed technologies suffer. Enterprises continue to use and produce technologies that cause damage to people and the planet because they are not held accountable for these damages which occur years or decades after they make massive profits from these harmful operations. To force the owners of these operations to be accountable requires identifying the current and potential damages that will result from production processes. The decision to move forward with these processes or avoid these processes must be a democratic decision based on scientific evidence of the consequences. This calls for a community wide educational effort. Communities and individuals must be made aware of the technology alternatives in all production processes and the real and potential consequences to the environment and human well-being.

In the dialectical relationship between historical materialism and the economic base, historical materialism is preeminent [5]. This places particular importance on the relations of production determined by the pricing of production and the competition among firms and capitalists. Firms compete for labor and markets. When market size increases, production increases, demanding more labor. This increase in labor demand empowers labor (workers) to demand more in wages, benefits, and control over production. Firms respond by engaging technology and techniques to increase productivity and decrease the dependence on labor. The production process engages a given proportion of different raw materials, labor hours, and equipment. Production results in either a fixed amount of consumer goods, or a given amount of equipment or processed materials that serve as input to other production processes. The relations of production established by capitalist competition promote an increased utilization of equipment and processed materials to increase worker productivity. At the same time, this increases the complexity of the supply chain, making production subject to instabilities. While production output increases is comes at the cost of a reduction in the proportion of the resources and profits allocated to workers.

The Forces of Production

A careful examination of the technologies added to the production process is important in determining the impact on the environment and people. The owners of production plot out the course of technology development with one primary goal, increasing their wealth. Owners are less concerned with the environmental impact of the technologies. In many cases, there is an intentional effort to hide information on the potential damage technology, techniques and end products will have on the environment or health and well-being of workers and consumers. Decisions regarding this significant part of the forces of production are in the hands of owners and their top managers. The pace, complexity and expense of technological change is accelerating. This increases the possibility and significance of technological damages. To offset this, a mandate to increase monitoring and control over technology advances is necessary.

The drive to increase productivity increases the power of the owners of production through a reduction in labor power required. Also, the increased value of the assets of a firm in its equipment and inventories of processed materials and raw materials makes competition more difficult. With the high cost of mechanization, few companies can compete. The increased wealth of the owners gives them more financial leverage through credit opportunities. Increased mechanization reduces the power of the workers by making them more dependent on the means of production and more expendable by reducing the need for workers. A more complex production process requires workers of different skill level creating further division among workers. This is a significance expression of the relations of production. Division among workers reduces their potential to influence the production process. The increased productivity works best when there is an expanding market. The increased productivity will primarily contribute to increased profits for owners. It can also lead to an increase in wages. These increased wages will allow increased consumption and drive increased production of consumer goods. Deploying more consumer goods and services is important in improving the well-being of citizens. Techniques and technologies promoted in the expanded delivery to consumers can be an important aspect of promoting appropriate technology. An imbalance in this production process due to, insufficient wages given to workers, unaccounted for damage to the environment, or too much focus on unproductive labor can result in a reduced consumer base and lead to overproduction. Overproduction can lead to a decline in production and result in cutback in labor. This further reduction in workers means further reduction in consumption and a downward spiral. This is the possibility that the World Economic Forum seeks to avoid in addressing their perspective advancing the 4th Industrial Revolution (4IR)

Those engaged in production are the most important part of the forces of production. They employ the techniques and equipment in the production process. Workers are provided with the machinery of production and given training on its use and maintenance. They develop skills in the production processes. Control over the technology, techniques and machines of production is the critical piece. These production elements (or forces) are developed to increase the profits of the owners of the means of production. However, the increased complexity of this machinery of production requires a sophisticated workforce. There is opportunity for workers to leverage their unique skills to demand more control over the production process. A highly technological workspace has been created where a few skilled workers are critical to production, while a wider number of unskilled workers are highly expendable. Skilled workers receive more money and benefits to ally with the owners and split from the larger mass of unskilled workers. These skilled workers are the users and promoters of the sophisticated technology. They are key in further research and development of these technologies. Some members of this skilled group are conscious of the damage technologies cause to the environment and community. With this knowledge a few step up as advocates for changes and whistle blowers exposing human and environmental injustices. Other skilled, knowledgeable workers turn a blind eye, accepting, and in some cases justifying the inequalities. This skilled group engaged in the production process alienates themselves from the bulk of workers. They realize their skill set makes them more essential to the production process. This confidence provides the opportunity to express concerns over the impact that these technologies and techniques have on fellow workers, the community and environment. Being the most knowledgeable of how this technology developed and operates, this set of workers is most aware of alternative options. They are most knowledgeable of appropriate technologies and techniques. However, the success of these skilled workers is linked to accepting and promoting the technology trajectory that the owners of production have decided upon. Alternative suggestions are only welcome when they bring more wealth and power to the owners of the production process.

Relations of Production Impact Forces of Production, and Direct Technology Development

Historical materialism can be applied at three levels. First at the level of the firm, factory, or organization. At this level, the owners have control over the production space, machinery, and tools. They make decisions on the types of technology used in production and decide on the products produced, prices and any restrictions on who can receive the products and services. These owners decide on the nature of the technology infused in the products and services. The second level is the national level. Owners compete and collaborate in decisions around the technology used in production and resulting from production. The state apparatus serves to mediate this process through state agencies, laws, technology policies and, in some cases, judicial decisions. The tension between owners (capitalists) and the workers and communities takes shape on the national landscape. However, in different nations, certain industries and major multinational corporations play an outsized role. These enterprises operate across supply chains and national boundaries. They can make a broader range of decisions regarding the technologies on the national scale.

The third level of operation is the international plane. Globalization has linked financial and production operations across nations. Historical materialism and the globalization of capital is particularly significant with advances in technology leading to complex global value chains and supply chains [6]. Multinational corporations and international organizations collaborate to determine the bigger picture regarding technology development and deployment. Historical materialism can be applied to connect capitalism to global wars and the global crisis generated by these wars. This concern must extend to using historical materialism to assess other forms of global aggression, such as sanctions, financial blockades, arms sales, military exercises, covert intervention, and cyberattacks. Global aggression sanctioned by multinational capitalism engages people in production with a range of technologies. There is an urgency for communities and people centered organizations to connect across national boundaries to counter the power of multinational capitalism. There must be a call to direct research, development, and deployment of technology in a direction more beneficial to the people and the planet. "Indeed, the single-minded goal of technological innovation under capitalism is expansion of production, profits, accumulation, and wealth for those at the top, not protection of the environment." [8, p. 44]. The ecological damage done to date is causing the spiraling forward of climate changes of rising temperatures, melting Arctic Sea ice, extinction of species, acidification of the ocean, and carbon in the atmosphere rising daily. An ecological revolution is needed to redirect technological development. This will come from the global South [8].

How can dialectics between forces of production improve relations of production and empower people through appropriate technology? One line of thought is that historical materialism can be applied to the management of enterprises to promote democratic engagement that can lead to socialist operation. Two possible scenarios were proposed that can lead to worker empowerment in the workspace [9]. The first scenario is the conversion of run-down capitalistic enterprises into cooperative firms. The second scenario requires the government to pass laws converting existing companies into worker owned enterprises. Both scenarios would require workers and communities to have the skills and tools to run the firms more effectively than the capitalists.

The capitalists would not sit idly and allow the seizure of their assets. The global operation of capitalism would result in capital flight from any country that attempted this transfer of power. Individual small firms may be converted to worker-controlled enterprises. However, global monopoly capitalism will not tolerate competition outside the realm of capitalist organizations. This does not rule out the significance of efforts to create and promote worker-controlled and community-controlled enterprises. These can be important cogs in the more comprehensive process of majority control of resources, production, distribution, and reproduction on the national level and within the global context. This will only come when communities and individual workers seize control of the skills, techniques and technologies of the production, distribution, and reproduction processes. With this control, they must direct the technology agenda away from the conquest mentality toward a focus on the needs of the masses and the well-being of the planet.

Education and skills are important components of the forces of production. Critical thinking must be reinforced as a skill used to further education. Critical thought applied to the situation of inequality should lead to organized efforts to direct technology to benefit a broader global community. These efforts will apply pressure to the current relations of production demanding a larger share of human, natural and financial resource be used to develop and deploy technologies to address basic needs of the most disadvantaged.

Skills development plays a particular role in reproduction of labor and therefore the development of the forces of production. The relations of production under capitalism places the control of educational institutions in the hands of the owners of production. These institutions focus on providing the training requested of the owners. This training is to advance capitalist plans for technology development and production. Capitalist plans focus on maximizing profits for the owners through minimizing the benefits to the non-owners and ignoring the current and future damage to the environment which these technologies and production cause.

The fundamentals of dialectical and historical materialism are the basis for refining system thinking that leads to developing systems dialectical thinking. System dialectics promotes the study of systems as a scientific subject and the application of system thinking in the study of all scientific disciplines. All scientific research should employ systematic and dialectical methods [10]. Jie promotes the grounding of the wide range of system science approaches in dialectical and historical materialism. Examples include cybernetics, informatics, operational research, differential dynamics systems and fuzzy mathematics. These are quantitative techniques that can be applied to multiple disciplines to address problems and assist in decision making.

Some Fundamentals of System Dynamics (SD)

The application of system dynamics as a paradigm is based on the system dynamics methodology developed by Jay Forrester at the Massachusetts Institute of Technology (MIT) [11]. System dynamics examines the causal relationships of key variables that force feedback relationships. Of particular concern is how positive and negative feedback relationships combine with delay to cause unanticipated behavior patterns. Conceptual models were used to capture the concepts of dynamic causal relations, delayed reactions, and both positive and negative endogenous feedback.

The fundamental building blocks of system dynamics models are stocks, flows, connectors, and auxiliary variables (convertors) as displayed in figure 2.1 below. A stock reflects a quantifiable variable that can change over time. In this example population is the Stock. Flows enter or leave a Stock. They reflect the change in a unit of time in the Stock. Population growth is an inflow contributing to an increase in Population. Convertors are constants or variables. Connectors are information flows that are the inputs to convertors that are variables and to Flows. The Connectors linking growth rate and Population to Population growth allow setting Population growth = growth rate * Population.

Figure 2.1: System dynamics fundamental building blocks

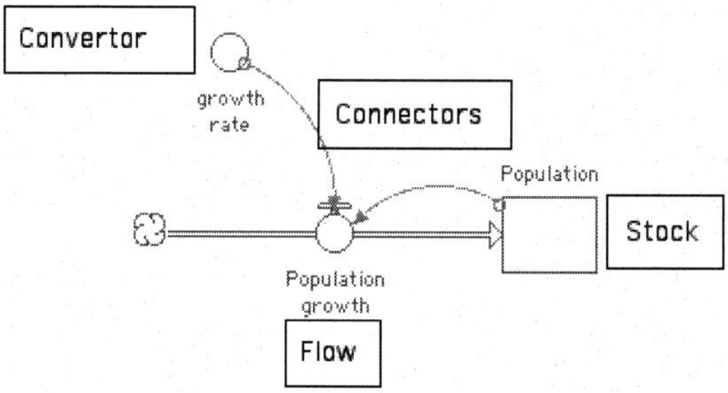

Forrester advanced the possibility of applying non-linear feedback theory used in engineering applications to social systems. Initially, the system dynamics methodology was applied to industry [11] [12]. He presented a technique for simulating the dynamics of industrial and economic activities. It includes conceptual and quantitative model development that can provide a range of computer simulation to aid in decision making. [11]. System dynamics research and applications quickly expanded to consider national and global problems

Figure 2.2: Impact of stress on growth [12]

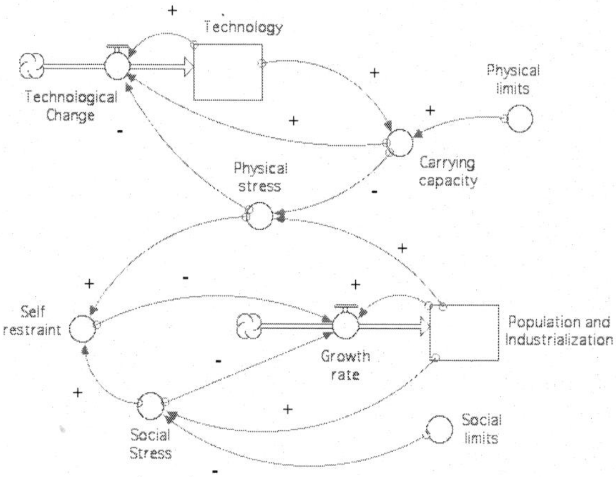

Figure 2.2 above reflects Forrester's early examination of technology and industrialization on a global scale. It is a highly aggregated representation of 'population and industrialization' and 'technology' interacting with physical and social stress. Each link in the diagram is labeled with + or – to indicate the polarity of the relationship. If 'Social stress' increases, 'Growth rate' will decrease. An increase in 'Social stress' contributes to an increase in 'Self restraint'. The increase in industrialization and population leads to an increase in both physical and social stress. In turn, the increase in stress slows growth and technology change.

While this research and applied work started at MIT, it has developed into a globally accepted field. Now system dynamics is employed globally in a growing set of circumstances [13] [14] [15] [16]. System dynamics is both a paradigm and a methodology. As a paradigm it is recognition that all systems are a complex of dynamic relationships that lead to feedback creating non-linearity. Also, the existence of delay in causal relationships leads to unanticipated consequences. As a methodology, system dynamics starts with identification of a concise problem, proceeds with model construction, and concludes with the testing of the impact of selected policies. Model construction is qualitative and quantitative and involves several steps.[2]

Expanding the concepts in Figure 2.2, system dynamics can be applied to the problem of 'technology development and deployment contributing to inequality'. The focus on endogenous feedback allows the study of how policy changes can create more favorable future patterns of behavior. This approach builds on the world models initiated by the Club of Rome studies and developed over the past 50 years [17], which facilitates a global, long-range examination of technology impacts. The effort is designed to lead to the development of meaningful quantitative models that can serve to assist policymakers in devising more effective technology policies. Club of Rome 'Limits to growth' report was a groundbreaking publication, with more than fourteen printings in multiple languages [17]. In this model, industrial output drives resource consumption and pollution generation, while at the same time providing the resources to produce services, agricultural capital, and additional industrial capital. This model aggregates population growth, industrial development, and pollution generation on a global basis. The original authors of 'Limits to growth' produced several books that extended the original effort [18], [19]. Randers models allow for the disaggregation along geographic lines and allows more discerning scenarios to be examined. None of these models make a distinction in technologies. Not all technology contributes to pollutants, just as technologies using renewable energy and other renewable resources have a different impact on resource consumption.

The World model developed by the MIT team for the Club of Rome focused on the impact of exponential growth under the constraints of limited resources. The exponential growth in world population and industrial development placed constraints on limited natural resources and limited space. At the same time resulting industrial pollution damages the planet and impacts human quality of life. Since the initial model, massive evidence points to serious consequences from damage already done by unfettered industrial growth. Since the initial model, significant evidence has been provided detailing the dynamic relationship between resource utilization causing pollutants like CO_2. This in turn has lead to rising global temperatures which lead to a rise in ocean levels.

These models need further development to distinguish industrial damage done by developed countries. This will point out the extent of damage attributed to the conquest mentality. More importantly, system dynamics models can illustrate how a focus on appropriate technologies to provide basic needs to the mass of disadvantaged and dispossessed will shift the imbalance of resources utilization and improve well-being for a broader population.

Currently much of the technology in place contributes to pollutants and other environmental degradation. Technology can be broadly categorized as less appropriate technology or appropriate technology (AT). AT technology contributes to reduction in pollutants, empowering of communities and ecological balance. Less appropriate technology contributes to pollution, environmental degradation, and inequality. Figure 2.3 illustrates the role of both types of technology. The increase in pollutants and other damage to the environment are motivations to increase the push for AT. This push for AT increases the proportion of new appropriate technologies in place of lesser appropriate technologies. The impact of this changing balance in technologies should reduce the pollutants completing the feedback loops between technologies and pollutants. Both categories of technology depreciate over time causing a demand for new technology. There is a delay related to this depreciation and a delay related to meeting the demand for new technologies.

Figure 2.3: Expanding Appropriate Technology (AT) or Less Appropriate Technology

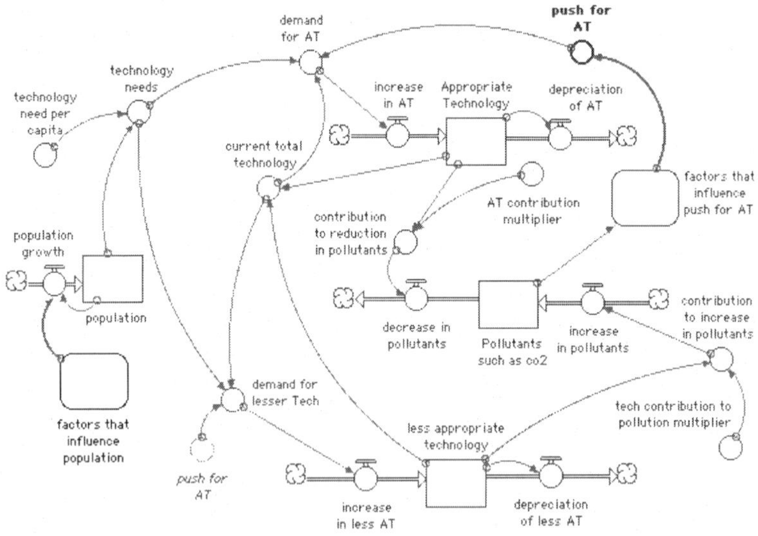

This model considers the population and population growth. It is appropriate for scenarios on national, regional, and global level. A similar model can be constructed at the firm level where the technology needs are determined by the Firm's strategic plan

instead of the population of the region or country. In all cases, the technology needs are satisfied by choosing between appropriate technology and less appropriate technology. An increase in pollutants motivates stakeholders to increase the push for AT. The 'push for AT' is a variable between zero and one and represents the proportion of technology that is appropriate technology. The proportion of less appropriate technology is [1 – 'push for AT']. Together the two categories fulfil the total demand for new technology. In this model pollutants are the only factor identified that impacts the push for AT. However, other factors must be considered that impact the push for AT and the pushback toward traditional less appropriate technologies in a more complete, robust model. Figure 2.3 represents a class of models where a push for AT can influence the balance in technology and the level of pollutants. In this case, AT is Technology development that is sustainable and collaborates with the planet's ecosystem. This conceptual model is useful in illustrating the dynamic relations among key factors in this scenario of pollution and technology decisions. Quantities and functions can be assigned to these variables using simulation tools such as Stella and Vensim[3]. Computer simulations can provide useful information projecting the future development of key variables. In this case the selected balance between appropriate technology (AT) and less appropriate technology (Bad Technology) can determine whether Pollutants increase or decrease.

The causal loop diagram (CLD) is a second conceptual model construct used in the system dynamics methodology. The CLD is used to highlight the feedback loops in the system as reflected in the model constructed. In Figure 2.4, there are two feedback loops. In the top loop – when Pollutants increase, there is a push for more AT decreasing the demand for less appropriate technology. This in turn decreases the deployment of bad technology leading to a decrease in pollutants. In the bottom loop, an increase in Pollutants leads to a push for more AT increasing the demand for appropriate technology. This increases the deployment of appropriate technology leading to a decrease in pollutants. Both are negative or balancing loops contributing to equilibrium.

Figure 2.4: Causal Loop Diagram (CLD) of Expanding AT model

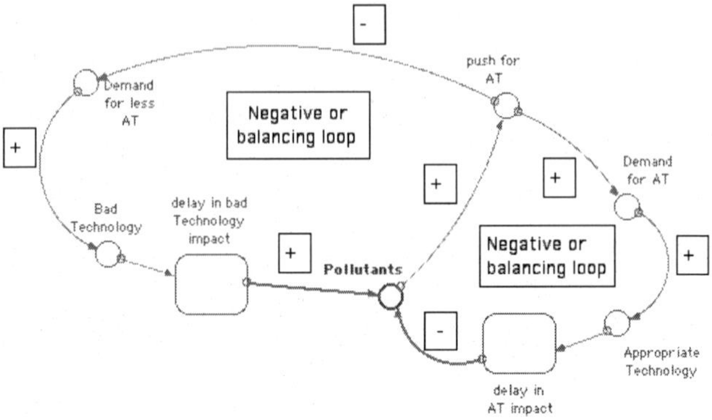

System dynamics can be used as a methodology to address a range of problems through constructing quantifiable simulation models. Table 2.1 represents possible considerations of technology focus on AT and the key stocks that impact technology selection. In turn technology choice impacts these key stocks, forming the major feedback process. These models allow the examination of various conceptual scenarios and computer simulations. This can provide the various stakeholders a better understanding of the technology options and their impact. The aim is the empowerment of community stakeholders. The public private partnership (PPP) model is seen as one approach to empowering communities and promoting sustainable people centered development. Government resources are provided through the public sector with an agreement from providers in the private sector to supply certain resources. System dynamics has been used to address associated scenarios using PPP models [20].

Conclusions

Systems thinking is an effective way of examining current technology dynamics and plotting a way forward that fosters a better understanding of the significant role of appropriate technology. The system thinking paradigm realizes that reality is composed of systems. A system is composed of elements linked through relationships. Historical materialism is a system thinking paradigm

Table 2.1: Appropriate Technology Options

Technology Focus	Key stocks
Technology development that is sustainable and collaborates with the planet's ecosystem	Pollutant such as CO_2
Technology for basic needs: food security; shelter; safe water and basic sanitation; and access to energy	Food supply; Housing; Water supply; Energy generation
Technology for basic services: healthcare; education; transportation; and communication.	Healthcare funding; Education funding; Transport funding; Communication funding
Technology development that utilizes local material and local labor	Material utilization; Labor utilization
Technology development that minimizes transport of goods and labor	Transport miles; Labor utilization
Technology development that minimizes unproductive labor, particularly that associated with warfare and violence.	Labor utilization; Military funding
Technology development that empowers local communities	Community resources; Labor utilization
Technology development that focuses on circular economy minimizing waste, particularly non-degradable and toxic waste.	Recycled technology; Waste

where the primary elements are the forces of production and the relations of production. In turn, technology and people are key elements in the forces of production. The impact that the forces of production have on the relations of production are the determining factor in societal development. People's technology choices determine their level of empowerment and the ecological impact on the planet. Historical materialism can be used to better understand how appropriate technology can empower communities and help restore our planet's balance. Extending dialectical thinking to non-linear feedback systems brings into focus the significance of system dynamics as a paradigm that encompasses historical materialism. Construction of stock and flow models and causal loop diagrams allow the sharing of various scenarios of system conceptualizations. Shared experiences can lead to a better collective understanding of how to address current and future technology related problems. Quantifying these models and running computer simulations strengthens the justification of these models and validates representations of our system reality and provides clearer decision alternatives regarding the development and deployment of technology.

Historical materialism identified a clear division in the interest of the owners of technology and the diverse populations engaged in developing and utilizing the wide range of technologies active in society. The identification of damages caused by extensive use of technologies should be a concern of all stakeholders - owners and users. It is necessary to examine closely how these stakeholders are organized and how advocates of appropriate technology can use tools and methodologies like historical materialism and system dynamics to encourage the range of stakeholders to increase their advocacy for appropriate technology.

Bibliography

[1] E. Williams, Capitalism and Slavery, New York: G. P. Putnam's Sons, 1980.

[2] I. Gollobin, Dialectical Materialism Its Laws, Categories, and Practice, New York: Petras Press, 1986.

[3] J. Stalin, Dialetical and Historical Materialism, Dickinson, ND: SAI Press, 2021.

[4] M. Cornforth, Historical Materialism, Second edition ed., New York: International Publishers, 1975.

[5] S. Amin, The Law of Value and Historical Materialism, New York: Monthly Review Press, 1978.

[6] M. Rupert and H. Smith, Historical Materialism and Globalisation: Essays on Continuity and Change, eBook Academic collection ed., Abingdon: Routledge (Warwick Studies in Globalisation), 2016.

[7] A. Bieler and A. D. Morton, Global Capitalism, Global War, Global Crisis, Cambridge, UK: Cambridge University Press, 2018.

[8] J. B. Foster, B. Clark and R. York, The Cological Rift Capitalism's War on the Earth, New York: Monthly Review Press, 2010.

[9] B. Jossa, "Is Historical Materialism a Determinisitic Approach? The Democratic Firm and the Transiton to Socialism," *Review of Radical Political Economics,* vol. 50, no. 1, pp. 82-98, 2018.

[10] W. Jie, Systems Dialectics, Beijing: Foreign Languages Press, 1996.

[11] J. Forrester, Industrial Dynamics, Cambridge, Massachusetts: MIT Press, 1961.

[12] J. Forrester, World Dynamics, 2nd edition ed., Cambridge: Wright-Allen Press, 1973.

[13] J. Sterman, Business Dynamics system thinking and modeling for a complex world, Boston: Irwin McGraw-Hill, 2000.

[14] J. M. Garcia, Theory and Practical Exercises of System Dynamics, Kindle ed., Juan Garcia, 2021.

[15] A. Ford, Modeling the Environment, Washington Dc: Island Press, 1999.

[16] P. Senge, The Fifth Discipline, New York: Currency doubleday, 1990.

[17] D. Meadows, D. Meadows, J. Randers and W. B. III, The Limits to Growth, New York: Universe Books, 1972.

[18] J. Randers, 2052 A global Forecast for the Next Forty Years, White River Junction, Vermont: Chelsea Green Publishing, 2012.

[19] D. Meadows, D. Meadows and J. Randers, Beyond the Limits: Congronting Global Collapse, Envisioning a Sustainable Future, White River Junction, Vermont: Chelsea Green Publishing, 1992.

[20] W. Liu, X. Wang and Q. Guo, "Impact of the collaboration mechanism of PPP projects based on sonsumer participation: A system dynamics model of tripartitie evolutionary game," *PLoS ONE,* vol. 16, no. 9, pp. 1-28, 2021.

[21] J. Harper, „National Defense Magazine," 2021. [Online]. Available: https://www.nationaldefensemagazine.org/articles/2021/5/28/pentagon-requesting-boost-in-rd-funding-to-compete-with-china. [Accessed 16 October 2021].

[22] P. Bajpai, "Nasdaq," 2021. [Online]. Available: https://www.nasdaq.com/articles/which-companies-spend-the-most-in-research-and-development-rd-2021-06-21. [Accessed 16 October 2021].

Endnotes

1 In the USA, "the Pentagon's fiscal year 2022 budget request, released May 28, calls for $112 billion for its research, development test and evaluation accounts. That would represent a $5.5 billion, or 5 percent, boost over the amount enacted for 2021. It is the largest RDT&E request in the history of the department. ... The department plans to spend over $66 billion on efforts related to the Indo-Pacific region, including for the Pacific Deterrence Initiative, also known as PDI. That includes $5.1 billion in targeted investments aimed at developing and procuring capabilities in support of 'joint force lethality,' including precision strike and stand-off systems such as advanced munitions. ... For 2022, the department is requesting $2.3 billion for microelectronics efforts, which are critical components for the Pentagon's weapons platforms. ... Meanwhile, about $3.8 billion would go toward hypersonics weapons. The Army, Navy and Air Force are pursuing ground-launched, sea-launched and air-launched versions, respectively. Artificial intelligence projects would receive $874 million ... That would help fund more than 600 AI-related efforts". [21].
Companies that spend the most on R&D: Amazon's SEC filing reveals an expenditure of $42.74 billion in fiscal 2020 (11.1% of its net sales), Alphabet spent $27.57 billion on R&D, which is equivalent to 15.1% of its revenue of $182.57 billion during the fiscal 2020. The company's R&D sending has more than doubled since the fiscal 2016. Huawei, one of the world's largest patent holders spent over $22 billion on R7D in 2020; the company was granted 2,761 patents in 2020, an increase of 14% vis-à-vis 2019. Overall, Huawei has more than 100,000 active patents across 40,000 plus families. The number of people working in R&D at Huawei is around 105,000, which is more than half of its total workforce. Microsoft spent $19.27 billion.

Apple and Samsung spent $18.75 billion each, while Facebook spent $18.45 billion [22]

2 There are differing versions of the steps in the system dynamics modelling process. The following is from [15]. The reference mode is the dynamic pattern of the major variables identified with the problem under study.
Step 1: Get Acquainted with the system
Step 2: Be specific about the dynamic problem
Step 3: Construct the stock and flow diagram
Step 4: Draw the causal loop diagram
Step 5: Estimate the parameter values
Step 6: Run the model to get the reference mode
Step 7: Conduct Sensitivity analysis
Step 8: Test the impact of policies

3 Stella and Vensim are icon-based simulation software that can be used on both PC and Mac computers. They allow a two-dimensional graphical representation of the system under study. Micro-world interactive simulations can be used to gain a better understanding of the problem under study. Both use graphs and tables to display simulation results in an interactive environment. Euler and Runge-Kutta numerical analysis techniques are used to approximate solutions to differential equations in both simulation programs.

Chapter 3

The Stakeholders in the Quest for Appropriate Technology

John Trimble

This chapter examines the role and potential of the various stakeholder groups. The manifesto is designed to impact technology by motivating the different stakeholders to serve as activists for appropriate technology. Different stakeholder groups play distinctly different roles in developing, deploying, and utilizing technology. Each group has different relationships to the development and implementation of technology policy. The audience for this manifesto can directly influence the different stakeholders. Hopefully, this manifesto will empower citizen activists to engage the different stakeholder groups in an organized, systemic manner to forge a collective approach to empowering people and communities through appropriate technology. It is the citizen activist that must hold all other stakeholder groups accountable to the people.

Introduction

Stakeholder theory and stakeholder management is generally addressed within the context of the firm or a single organization. It is used as a management tool to assure the smooth running of the firm. A particular subset of Management is tasked to track the concerns of management and other stakeholders and makes efforts to address these concerns while maximizing the gains of owners and management.

> "The basic idea of 'managing for stakeholders' or 'value creation stake-holder theory' is quite simple. Business can be understood as a set of value-creating relationships

> among groups that have a stake in the activities that make up the business. Business is about how customers, suppliers, employees, financiers (stockholders, bondholders, banks, etc.), communities, and managers interact and create value. To understand a business is to know how these relationships work" [1]

The effort of this manifesto is to motivate stakeholders to operate for the welfare of global society by promoting appropriate technology. 'Value-creating relationships' can be created and nurtured among the stakeholders engaged in technology decisions. Fundamentals of stakeholder theory must be applied at both a local organizational and firm context all the way to the international arena.

The major stakeholder groups addressed throughout this manifesto are: 1) educational institutions; 2) government structures; 3) international agencies; 4) non-governmental organizations (NGOs), non-profit organizations (NPOs), and community organizations; 5) the private sector; and 6) engaged citizens.

Educational institutions play the most significant role in 1) research of technologies, 2) teaching of science and technology and 3) training of scientists, engineers, and technologists. Curriculum development and research agendas become determinants in the trajectory of technology development and deployment.

Government structures are responsible for delivering services to its nation's citizens and assuring that basic human rights are met. International agencies are extensions of these national governments, also responsible for assuring human rights associated with basic needs such as water and sanitation, education, food, and shelter. Non-governmental organizations (NGO), community organizations, and non-profit organizations all claim agendas designed to help meet the needs of people. To achieve this mandate, it is essential they understand the role of alternative technologies.

Powerful international agencies like the United Nations (UN), International Monetary Fund (IMF) and North Atlantic Treaty

Organization (NATO) play significant roles in setting technology policy and deploying various technologies. International organizations sometimes have equal representation per country. However, more often the more powerful nations have more influence in the operation of these international organization. Nations send representative to these organizations to serve the interest of their respective countries. Globalization has increased the role these organizations play in determining international policy.

NGOs, non-profit and community organizations consist of a wide variety of organizations. NGOs are generally non-profit organizations that deal on the international or national level with large projects addressing concerns that generally full under the purview of the government, such as public health, energy, and education. Non-profit organizations (NPOs) generally operate on a local community level with smaller projects often focusing on areas not addressed by the government. Community organizations could be non-profit, or profit based and often draw support from larger and more established NGOs, NPOs and international agencies. Community organizations usually have a particular thematic focus and specific geographic location.

Globally, the private sector, from small and medium enterprises to multi-national corporations, is responsible for most production and distribution of goods and services. Private ownership of these enterprises places the world's most substantial financial and material resources in the hands of a small number of global citizens. They are positioned to have the most significant impact on the implementation of technology policy. These owners are motivated to seek profit but are not monolithic in their ambitions. Firms provide a very diverse range of services and products, utilize varied natural resources, and impact the ecosystem in different ways. Different chapters of this manifesto will highlight some of the differences between firms and focus on firms relevant to the technologies addressed in each chapter. An ethical appeal will be made to owners, as key stakeholders, to support the positioning of appropriate technologies throughout their operations. A case will be made for citizen activists to hold these owners accountable.

The ultimate benefactor of technology policy is the citizen activist. The activists also have the potential for having the most influence on other stakeholders in the establishment and implementation of technology policy. This can have the most serious impact on the further development of appropriate technology.

Historical Materialism, Class Struggle and the People's Class

In the previous chapter we highlighted the significance of historical materialism in understanding the production process and the role of technology in advancing this process. Technology is the key aspect of the forces of production in empowering people and communities. The divide being the owners of significant wealth and the masses of poorly paid workers and unemployed is an antagonistic divide in communities. The wealthy capitalist class is engaged in daily struggle with workers and other disempowered and oppressed classes. The struggle is over the control of production, and where the products and profits from production go. The capitalists currently dominate in this struggle and use their upper hand to increase their profits and increase their control. They use their ownership, their wealth, their control over institutions to force others to work for the capitalist interest. It is obvious that we must empower workers and oppressed people to promote a more democratic and just society. This requires a broader mass control over the direction of technology development and implementation. Class struggle between wealthy profiteers and masses of underprivileged and oppressed requires a commitment to the interest of the masses. This practice of working for the interest of oppressed people and communities, places one in the 'Peoples' class [2]. While the relationship between the capitalists and the oppressed is the main tension between stakeholders in the production process, we choose to take a more nuanced approach to examining stakeholders and their role in reducing inequality through advocating appropriate technology. This does not deny class struggle. Instead, it considers the complex interactions of institutions and interest groups in the production process. This is consistent with many social scientists that elaborate on the significance of different sectors of civil society in the interactions of production and cultural engagement. Democratic institutions and processes are stressed in examining class interaction [3].

Intellectuals can be considered as a special group that exist withing the different classes and is the subgroup of all classes most engaged in the battle of ideas [4]. With that in mind, the manifesto intends to appeal to the critical thinking of all intellectuals regarding the necessity of appropriate technology. Gramsci also contributes to theory on civil society. One of Gramsci's contributions was an understanding of the development of consent and submission and how its reproduction in the milieu of advancing capitalism impacts democracy [5]. This manifesto serves as an appeal to the intellectuals of all stakeholder groups. An appeal to reasoning and ethics. Do not submit to the demands of the few, instead support the democracy of the full population impacted by technology. The following sections provide a guide to addressing each stakeholder sector.

Many popular writings on historical materialism group all structures and institutions outside of the economic realm into the superstructure. This groups together the governing structures of the state, educational institutions, religious organizations, and other elements of civil society. When capitalism emerged the state and religious institutions were closely linked, while educational institutions and other civil society structures were very limited.

Civil Society

Gramsci was one of the first intellectuals to recognize the development of civil society organizations (CSOs). He linked this to the significance of the intelligentsia and the role of political parties in civil society [4]. Under capitalism, because capitalists dominate in the realm of ideas, civil society is inevitably "a place of formation and reproduction of bourgeois values, it is the first stage in the competition of the social classes ... Civil society thus appears as a place in which the relations of state and economy, private and public sphere are redefined, under the terms of ideological competition" [5, p. 1]. Gramsci saw the superstructure as divided into civil society and political society. Political society (the state) used the coercive power of laws, courts, and police to enforce the will of the dominant class. The dominant class has hegemony in civil society and uses its ideological dominance to achieve the consent from the masses [4]. In recognition of the increased significance of civil society, political society has shifted attention from government to governance. In this context "three dimensions of the relationship between government and CSOs are presented" [6, p. 25]. They are: 1) ideological – advances different worldviews and political strategies; 2) institutional – focus on laws, informal practices, and communication; and 3) street level – develops strategies dealing with implementation of official policies.

The World Economic Forum (WEF) and its founder, Klaus Schwab, have placed considerable attention on defined civil society and examining its future role. Defined as the space outside the family, market and state, civil society includes NGOs, nonprofit organizations; Online groups and activities including social media; Social movements of collective action; Religious leaders and faith-based organizations; Labor organizations; Social entrepreneurs; Local grassroots organizations and activities; and Democratic member-controlled cooperative [7]. This report identifies the driving forces the WEF feels will reshape civil society. The report presents four alternative futures: Mad Max; Transparently Blurred; Turbulence and Trust Deficits and Privatized World. None of these scenarios reflect the empowerment of masses. Mad Max is characterized by international and national conflict and strong government control. The Transparently Blurred scenario has government and the private sector addressing societal challenges. Turbulence and Trust Deficits is a chaotic where lack of trust leads to more local focus. In the Privatized world scenario governments have failed, inequality is extreme, and corporations dominate. The presentation of these as the only alternatives is how the WEF promotes its hegemony. The definitions and discussion projected by WEF becomes the starting point for other researchers and intellectuals engaged in presenting the role of civil society. Civil society's role in Bangladesh and Zambia is presented as examples of the global south [8]. However, the report is grounded in the thinking of the WEF. This report starts with the WEF definition of civil society and follows with over thirty citations of the 2013 WEF report. It serves as an example of the extension of WEF intellectual hegemony. WEF is not the only civil society organization competing for hegemony in the CSO space. UNESCO presented a report on the future of CSOs [9]. The report concludes that CSOs face a shrinking space and makes several recommendations including the need to serve as "the voice for diversity, inclusion and collective thinking ... support the development of free will, critical thinking education and learning at community level to foster active and empowered citizens equipped with relevant skills for the future" [9, p. 12]. What is evident in all these positions on CSO is the increased integration of civil society with the state and economy. The dominant class increases the complexity of its engagement to maintain and increase control. In a climate dominated by infiltration of civil society by the state and capitalists, there is an argument that people centered CSOs can influence the state and economy without compromising their agenda [10].

The positions of WEF and UNESCO regarding CSOs is indication of the special role international agencies can play in the CSO landscape and reason for examining them as a separate stakeholder community. The massification of education in the 20th century and the current claim that education is a human right is the basic for distinguishing educational institutions from the rest of civil society. This manifesto places a particular significance on the activist citizen as the most critical stakeholder in advancing appropriate technology. Citizens must use activism to mandate all other sectors to promote appropriate technology.

Educational Institutions

In much of the world primary and secondary education is a standard. The world literacy rate has gone from 21% in 1900 to 86% [11]. This is largely due to the institutionalization of education in most countries. The educational agenda at the primary and secondary level impacts the workforce as well as the substance of university education and research. This agenda sets the standard and provides the qualified students at the tertiary level.

> "There are close to 85 million teachers worldwide: 9.4 million in pre-primary; 30.3 million in primary; 18.1 in lower secondary; 14.0 in upper secondary; and 12.5 in tertiary education. By 2030, an additional 68.8 million teachers will need to be recruited just for primary and secondary: 20 million are required to expand access to primary and secondary school and 49 million are needed to replace those who leave the workforce" [12].

In developed countries teachers reflect a higher percentage of the workforce in less developed countries. However, in the global south, teachers role as leading intellectuals is more significant that their numbers. "Today, there are around 220 million tertiary education students in the world, up from 100 million in 2000" [13]. In the USA,

> "The number of colleges and universities grew from 1,851 in 1950 to 3,535 in 1990 to 6,900 in 2013. In the 1949-1950 school year 2.66 million students were enrolled in colleges and universities; by the 1989-1990 school year 13.54 million students were enrolled. In fall of 2013, 19.9 million students were enrolled in colleges and universities" [14].

These numbers of students and universities speak to the increased importance higher education institutions play in formulating ideas and shaping future intellectuals, scientists, and engineers. Everyone should be concerned with what is taught, as well as the research conducted at these institutions.[1] This manifesto is directed at students, academic staff, administrative staff, workers, and alumni of our universities. They can all play a role in reshaping the technology agenda of higher education. We, all must be concerned with the universities promoting appropriate technologies to empower people and their communities.

Government Structures

The nation is the primary determinant of laws and policies that govern the behavior of its citizens, corporations, agencies, and institutions. Each country's governing administrative bodies set and implement technology policies. National judicial systems interpret policies, enforce policies, and penalize policy violators across the country. Provincial, state or counties have governing bodies that determine and apply policies on various regional levels inside the country. Often these units play a more important role than national bodies. Governing structures at this level are generally responsible for roads, and providing electricity, water, and sanitation.

State owned enterprises play a unique role in production and distribution. They are economic entities that are directly controlled by government. There primary goal is not maximization of profits but meeting the needs of the people. In most countries postal services and water and sanitation facilities are owned by the state. Many countries own airlines and electric power stations. In many countries there a government departments or ministries responsible for coordinating state owned enterprises[2]. They should be held accountable to the interest of the people. Technology choices must be monitored. Governments on city and national levels can plan and implement technologies across enterprises in a manner that benefits and empowers communities particularly the most disadvantaged. This can be a coordinated effort to address inequality. Voters and their communities must hold the elected officials in charge of state-owned enterprises answerable to the people.

Municipal and city governing bodies have become more important as urban growth has escalated. Countries with more than 50% of their population in cities realize the significant role city government plays. Service delivery, transportation, pollution, jobs, and housing issues require the attention of city government, their planners, and regulators.

The responsibility for technology policies that would guide science and technology research through deployment lies with the government. In democratic societies where some version of one-person one vote operates it is important to hold government accountable. The creation and implementation of technology policies that promotes technology for the people is a demand that democracies must make. There needs to be a demand that government representatives create structures and policies that allow every citizen to become more engaged in the governance process. This itself suggest a form of appropriate techniques and technology.

International Agencies

Advances in transportation and communication technology has contributed to globalization. The increased role of finance, credit, and currency exchange also plays a significant part in globalization. With advances in globalization, international agencies from the United Nations (UN), World Trade Organization (WTO), International Monetary Fund (IMF), World Health Organization (WHO) and the World Bank have grown, in size and significance. These international organizations are major funding sources for a wide range of projects[3]. The project selection process must be scrutinized, carefully. How can project selection and implementation engage technology that will reduce inequality and improve ecological balance? These agencies engage in studies and generate reports directed to establish and support certain technology policies. Pressure must exist to influence these studies and critically examine these reports. Only then, can these international entities be held accountable and pushed to address appropriate technologies and techniques.

Regional multi-country organizations such as the European Union, African Union, BRICS, NATO, The Commonwealth, and the Africa Development Bank work to address concerns of a regional and global nature. The Human Resources, Science & Technology Department (HRST) department of the AU could be a leading organ in fostering appropriate technology across Africa. This department and other units of the AU could coordinate a continental effort to promote technologies to address basic needs such as food, water and sanitation, energy delivery and housing.

All regional organizations have a particular role in strengthening regional agendas. It is important that these regional agendas address inequalities within the region and those outside the region that are a result of the regional agenda. All regional organizations must be held accountable, not only by their constituencies but by the global citizenry.

Non-Governmental Organizations (NGOs), Non-Profits & Community Organisations

This group of stakeholders consist of a wide variety of organizations. Perhaps the oldest subgroup in this category are religious organizations. These are among the first organized efforts to address human rights and can play a particular role in advocating appropriate technology. Faith-based organizations are engaged in supporting education, and service delivery through various structures including faith-based universities. An ethical appeal to religious organizations must be made for their support of technologies that improve the well-being of the most disadvantaged people and communities.

NGOs are generally non-profit organizations that deal on the international or national level with large projects addressing concerns that generally full under the purview of the government, such as public health, energy, and education. NGOs are considered in a separate sub-category from non-profits that have a limited size and scope. Most Non-profit organizations (NPOs) operate on a local community level with smaller projects often focusing on areas not addressed by the government. Non-profits with a particular focus on basic needs, education and technology can assist in promoting appropriate technology[4].

Community organizations could be non-profit, or profit based and often draw support from larger and more established NGOs, NPOs and international agencies. Community organizations usually have a particular thematic focus and/or geographic community. The largest number of organizations fall in this category. Many are temporary that spring up to address short term immediate problems like incidents of police misconduct or lack of service delivery. Community organizations span a spectrum from book clubs, to block clubs to neighborhood watch teams to Bridge clubs. Many of these function for social and cultural purposes and don't recognize their potential in addressing inequality and improving community well-being. However, as an integral component of all communities, they must be challenged to play a role addressing appropriate technologies to empower their communities.

Private Sector

The private sector is the cornerstone of economic production under capitalism. Capitalism is globally dominant and is intent on expanding the private sector. Thousands of new enterprises spring up daily. There is an ongoing effort to privatize state owned enterprises. More and more wealth is being placed in the hands of private owners. The private sector contains millions of small businesses, often with single owner/employee, other times family run. At the other end of the spectrum are multi-national public stock companies that operate in hundreds of countries like Amazon, Apple and Coca Cola[5].

Governments and universities around the world have set up incubators and special facilities to support innovation, new business creation and ramp up for small businesses. For many engaged in this process, this is an effort to carve out a niche in the existing global value chain. It is more important to reshape this global value chain by directing innovation and technology policy toward appropriate technology. Currently much of the start-up drive is directed to filling places in the supply chain determined by large multi-national corporations. The technologies used are largely determined by existing production and multi-national corporate interest. The start-up community must be motivated by a principle of socially relevant technology and innovation that empowers local communities. This requires individuals to rethink their motivation. Governments and universities also must examine their role and direct the drive for new enterprises toward the utilization of appropriate technologies.

Multinational corporations (MNCs) with billions if not trillions of dollars in assets are responsible for development and deployment of most technologies. The largest of these companies employ hundreds of thousands across multiple continents. Their technology impacts millions of people. MNCs massive use of energy and water and generation of waste leave a clear ecological footprint on the planet. In recent years, there is a drive to hold MNCs socially accountable. The carbon released from corporate production has become a popular issue. The news regularly reports on MNCs asserting that they are moving to carbon neutrality. Corporate public relations offices are alleging there operations are improving the well-being of the community. These assertions and claims must not be taken on face value. They must be challenged and MNCs must be pushed to do more and to do better.

Citizen Activists

This manifesto is intended for all stakeholder categories. However, it is the citizen activist that can most effectively promote appropriate technology. This category of stakeholders must press all other categories to act on behalf of the most disadvantaged and engage appropriate technologies to reduce the gap between the advantaged and disadvantaged. At the same time, the activist must promote a consciousness of the ecological and human health damage done by many current technologies. Consciousness can be the driving force in pushing back from these damaging technologies. A significant proportion of damaging technologies exist with the growing military arsenals of the major capitalist countries. A call for disarmament and global peace must be high on the agenda of all citizen activists.

Citizens can make meaningful impact on the technology landscape through conscientious decisions as consumers and investors. As consumers, the choice of products to buy and services to support should be based on the engagement in appropriate technologies. A preference for recycled products can help. Investment decisions should consider the impact the investment has on empowering people and nurturing the planet's health. A clear choice is investing in alternative energy services over coal-based industries. In following chapters, particular categories of appropriate technology will provide guidance for citizen activists to make better individual decisions. Citizens are encouraged to make individual daily decisions that support the appropriate technology community empowerment schema. However, it is most important for citizen activists to engage in an organized collective fashion. Citizens most profound impact will come through organization at the community, regional, national, and international level.

Conclusions

A manifesto on appropriate technology must point out the flaws in past and current technology policy and utilization. More importantly, it must provide a guide for change. This guide will identify the technologies that are most beneficial to human well-being, global ecological balance and community empowerment. To effectively bring about change, strategy for change must be directed to potential change agents. With this in mind, this chapter identifies the different stakeholder categories that must address appropriate technology. Justification is given for the particular categorization of stakeholders. The contrasting roles and responsibilities of the different categories are highlighted. Citizen activists are identified at the critical stakeholder category that must link to all other stakeholder categories and serve as a catalyst for change.

Bibliography

[1] D. Wasieleski and J. Weber, Stakeholder Management, Emerald Publishing Limited, 2017.
[2] A. S. Toure, Africa on the Move, London: Panaf Books, 1979.
[3] E. M. Wood, Democracy Against Capitalism Rnewing Historical Materialism, 5th ed., London: Verso, 2016.
[4] A. Gramsci, "Selections from the Prison Notebooks of Antonio Gramsci," ElecBook, London, 1971.
[5] V. Maglaras, "Consent and Submission: Aspects of Gramsci's Theoryof the Political and Civil Society," Sage Open, 2013.
[6] R. Pauly, F. D. Rynck and B. Verschuere, "The relationship between government and civil society. A neo-Gramscian framework for analysis," Ghent University, Department of Public Governance, Management and Finance, Ghent, 2016.
[7] K. Schwab, "The Future Role of Civil Society," World Economic Forum, Davos, 2013.

[8] R. Cooper, "What is Civil Society, tis role and value in 2018?," Helpdesk report UK Department for International Development, London, 2018.

[9] UNESCO, "The role of Civil Society Organisations in 2050 and beyond," UNESCO, , 2021.

[10] S. Klein and C.-S. Lee, "Toward a Dynamic Theory of Civil Soceity: The Politics of Forward and Backward Infiltration," *Sociological Theory,* vol. 37, no. 1, pp. 62-88, 2019.

[11] wikipedia, "List of countries by literacy rate," 2021. [Online]. Available: https://en.wikipedia.org/wiki/List_of_countries_by_literacy_rate. [Accessed 8 December 2021].

[12] E. Molina, "The World Bank," 2021. [Online]. Available: https://www.worldbank.org/en/topic/teachers#1. [Accessed 8 December 2021].

[13] N. Arnhold, "The World Bank," 2021. [Online]. Available: https://www.worldbank.org/en/topic/tertiaryeducation#1. [Accessed 8 December 2021].

[14] ProCon.org, "Britannica," 2021. [Online]. Available: https://college-education.procon.org/history-of-college-education/. [Accessed 8 December 2021].

[15] E. Duffin, "Statista," 26 February 2021. [Online]. Available: https://www.statista.com/topics/829/college-and-university/#dossierKeyfigures. [Accessed 14 December 2021].

[16] S. Writer, "BUSINESSTECH," 1 November 2019. [Online]. Available: https://businesstech.co.za/news/business/350689/this-is-how-many-people-south-africas-biggest-state-owned-companies-employ/. [Accessed 14 December 2021].

[17] M. Büge, M. Egeland, P. Kowalski and M. Sztajerowska, "World Economic Forum," 2 May 2013. [Online]. Available: https://www.weforum.org/agenda/2013/05/state-owned-enterprises-in-the-global-economy/. [Accessed 14 December 2021].

[18] "Federal Foreign Office Germany," [Online]. Available: https://www.auswaertiges-amt.de/en/aussenpolitik/internationale-organisationen/vereintenationen/-/281336. [Accessed 14 December 2021].

[19] "IMF Factsheet," 3 March 2021. [Online]. Available: https://www.imf.org/en/About/Factsheets/IMF-at-a-Glance. [Accessed 14 December 2021].

[20] N. Levin, "Largest Non-profits in the World," 5 March 2019. [Online]. Available: https://largest.org/misc/nonprofits/. [Accessed 14 December 2021].
[21] "Wikipedia - List of largest employers," 10 December 2021. [Online]. Available: https://en.wikipedia.org/wiki/List_of_largest_employers. [Accessed 14 December 2021].
[22] M. Urosevic, "Largest Public Companies on Earth," 8 November 2021. [Online]. Available: https://spendmenot.com/blog/largest-public-companies/. [Accessed 14 December 2021].

Endnotes

1 This table indicates the scope of funding & expenditures at universities in the USA in 2019 [15]

	Funding
Federal funds for postsecondary education programs in the United States in 2019, by government department	74.8 bn USD
Funding received by research programs at universities from Department of Health and Human Services	20.9bn USD
Expenditures of Public Universities and Colleges	395 bn USD
Expenditures of Private Universities and Colleges	225 bn USD

2 South Africa has approximately 700 SOEs, below is a table of the largest five [16]

Company Name	Segment	Number of Employees
Transnet	Transport	56 718
Eskom	Energy	45 982
South African Airways	Airline	5 256
Denel	Defence	3 438
Safcol	Forestry	2 396

China has more SOEs than any other country [17]
"To estimate which countries have the highest state-owned enterprise presence among their top firms and to ensure comparability, we use the equally weighted average of shares of state-owned enterprises in sales, assets and market value of the country's top ten firms. The top five countries are China

(96%), the United Arab Emirates (88%), Russia (81%), Indonesia (69%), and Malaysia (68%)."

3 Funding of international agencies
the 2018/2019 regular biennium for the United Nations was approved by the 73rd session of the General Assembly on 22 December 2018 for 5.811 billion US dollars [18].
The IMF was founded in 1944 and has a membership of 190 countries. According to its website it can lend $1 trillion USD to its members and has a staff of 2,700 from 150 countries [19]

4 This table shows the largest non-profit foundations [20]

Name of Non-profit	Headquarters	Endowment
Bill & Melinda Gates Foundation	Seattle, WA, USA	$50.7 billion
Wellcome Trust	London, UK	$33.7 billion
Howard Hughes Medical Institute	Chevy Chase, UD, USA	$22.6 billion
Garfield Weston Foundation	London, UK	$12.6 billion
Ford Foundation	New York, USA	$12.4 billion

The largest of these, the Bill & Melinda Gates Foundation advocates education, health and sustainability and donates about $ 1 billion annually.

5 Walmart employs 2.3 million people. McDonalds employs 1.9 million and Amazon 1.5 million [21].
Using market capitalisation as the metric, the following table lists the largest public stock companies. All but Saudi Aramco and Tesla are Information Technology related companies [22]

Company	Location	Market Value
Apple	USA	$2.27 trillion USD
Saudi Aramco	Saudi Arabia	$2.03 trillion USD
Microsoft	USA	$1.81 trillion USD
Amazon	USA	$1.70 trillion USD
Alphabet	USA	$1.30 trillion USD
Tencent Holding	China	$896 billion USD
Tesla	USA	$827 billion USD
Facebook	USA	$761 billion USD
Alibaba Group	China	$725 billion USD

Chapter 4

AT and Services

John Trimble

The service sector has increased in size, significance, and technological sophistication over time. Basic needs such as education, water, sanitation, housing, and energy were once a family or individual responsibility. Today, they are considered basic human rights. National governments, civil society organizations, and international agencies have taken on the responsibility to assist in the delivery of services. The private sector has organized a significant network of enterprises that profit by providing services to communities, governments, and individuals. Today, in many countries, employment in the service sector exceeds the combined employment in agriculture, mining, and manufacturing. In all the technologically advanced countries this is the case. Urbanization has increased population density, complicating the delivery of services. Advances in technologies associated with the Fourth Industrial revolution increase the significance of communication services, as well as the role of information and communication technology (ICT) in the delivery of other services.

This chapter takes a close look at the changing roles of family, government, small enterprises, national and multi-national corporations in providing services. Services examined include education, roads and transportation, water and sanitation, energy, housing, communication, sports, and entertainment. Consumer services, beyond basic needs, are becoming significantly more important. Sports, entertainment, healthy living, and cultural services are contributing to an increase in human well-being and providing a much-needed job stream. It is important to consider how to engage newer consumer services to empower communities. Short range and long-term technologies and techniques for service delivery are considered. The chapter considers how to select and advance technologies and techniques that most effectively empower individuals and their communities.

The historical role and current responsibilities of the different stakeholder sectors (educational institutions; government structures; international agencies; NGOs and non-profit organizations; private sector; engaged citizens) are presented. Guidelines are presented on how each sector should address appropriate technology options to impact service delivery to empower communities and increase citizen well-being.

Early Historical role of Individual, Family, and Community in Providing Services

For centuries most basic services were provided on the family or small community basis. Education was the responsibility of the family. Energy used to cook or heat as well as water and sanitation facilities were provided by the family or local community. These services used limited and primitive technologies. Yet, the technology was in the hands of the individual, family, or local community. This places particular forces of production in the hands of the people and empowered communities. Families and the local community lost control over many services as new technologies developed and urbanization created new dynamics. In cities the provision of energy and waste disposal required more complex systems that serviced thousands if not millions of people. The advances in transportation technology combined with the increased density in populations to require more complicated transportation solutions combining rail, road and water systems. The state and private enterprises came to control services ranging from water and sanitation to energy, education, and transportation.

Technology Shifted Responsibility of Service Delivery to Major Enterprises and Government

The 20th century saw electricity become the major power source for most advanced countries. National grids were developed that distributed electricity across vast distances. This required massive generation stations. These expensive national projects were undertaken by national governments, directly or more often by commissioning major multinational corporations. As highway systems became increasingly essential to long distance travel, road construction became a more extensive operation requiring more equipment and financing. National governments, aided by large construction companies, turned to manage transportation systems on the national level. The growth of cities led to more complex and expensive transportation systems, as well as water and sanitation operations that had to handle in many cases millions of people. This scaling up of service delivery systems increased the role of research, development and innovation in technologies and techniques for the creation and delivery of services. It also came at the expense of reducing local community control over services. This scaling up has led to depersonalization where people are viewed as customers and consumers. Recent research on the theoretical and ethical framework for universal basic services has led to a better understanding of shifting our thinking from "customers and consumers to residents and citizens" [1, p. 541]. Efforts to enforce an ethos of appropriate technology to empower people and communities will contribute to this much needed shift in thinking and practice.

Alienation When Citizens and Communities are not Engaged with Service Delivery

When local communities no longer control their service delivery several problems can occur. National or regional control of service management and service delivery can lead to an increase in inequality. Service development and delivery tend to favor the sector of the population that has more wealth and power. Most of the population recognizes that their service needs are not met and can see how the service sector more effectively addresses the needs and desires of a smaller select population. Service developers and providers are not giving proper attention to the population most in need of services. Those responsible for service delivery are viewed as outsiders or even foreigners. An alienation between

Table 1: Percentage of workers by sector

Country or region	Year reported	Agriculture	Industry	Service
South Africa [2]	2019	5.28%	22.31%	72.41%
USA [3]	2020	1.5%	13%	80%
UK [4]	2019	1.05%	18.12%	80.83%
Africa [5]	2020	48.67%	13.26%	38.07%
Eastern Asia [5]	2020	22.63%	27.53%	49.84%

the community and service delivery institutions can develop. In many cases, the situation can erupt into violence. Over the past 15 years, the Institute for Security Studies Protest and Public Violence Monitor in South Africa has tracked thousands of protest incidences. "South Africa recorded more than 900 service delivery protests in six months from August to January, according to Minister of Police Bheki Cele, as the pandemic did little to halt them" [1]. Across the world, we find local politicians campaigns promising improvements in service delivery. Many elected officials have lost their job because of citizens dissatisfied with service delivery. Community based efforts to break this alienation have ranged from local repair of potholes to demands for the state so subsidize the cost of water, energy and transportation. When actively engaged communities see their demands met and achieve more control over service delivery, only then will the alienation reduce.

The Power of Service Workers: Current Organization and Potential

Table 1 above shows that a significant proportion of the workforce is service workers. While in Africa the agriculture sector is the largest part of the workforce, clearly the global trend, even in Africa is for the service sector to increase with technological and urban advancement. In cases like South Africa, UK, and the USA where the service workforce is 72% to 81% of the total workforce, this sector can control the direction of society's advancement if properly organized.

The difficulty lies in the diversity of this sector. It includes health workers, educators, government employees and police and private security guards. Different sectors of service workers have very different interest and priorities. They also have very different engagements and concerns with technology. What a city police force may see as appropriate technology to empower them is very different from the technology advocated by health workers or high school teachers.

The power of service workers must be the concern of all stakeholders. A significant amount of service provisions is paid by citizens through their taxes. Public libraries to military bases in foreign lands are paid for by the people. It is essential that everyone understand the role of these services and service workers. How do the different services contribute to the well-being of the population? How can an increase or decrease in resources to different services impact community empowerment and environmental security? This manifesto advocates technologies to empower all communities. However, the most disadvantaged communities must be the focus. Therefore, taxes and private sector funding should be directed toward basic needs services. To accomplish this a shift away from the trillions of dollars globally spent, each year, on military and policing technologies is mandatory. Out of self-interest, service workers and their families are motivated to demand more resources are put into service provisions and delivery. The drive for better services is too often narrowly focused on specific communities and services that have a louder voice in the government.

Services Workers Engagement with and Selection of Appropriate Technologies

Service workers need to be better educated on the impact that technologies have in their workspace. The service sector uses a significant amount of infrastructure. Just think – Post offices, police stations, railroad stations, trains, buses, libraries, schools, hospitals, clinics, theatres, and sports stations. Currently, workers have little say in the construction and maintenance of these structures. Measures addressing appropriate technology could include local materials and labor to improve building insulation, renewable energy on buildings using solar panels and wind turbines, rainwater capture on rooftops and recycling of water, electric vehicles, and the use of green hydrogen in buildings and vehicles. The rethinking of technologies in buildings and vehicles is the broadest approach to engaging appropriate technology across the service sector. Despite the diversity in services provided, the common use of buildings and vehicles is grounds for service workers collaborating on appropriate technologies. Labor unions and other workers' organizations must accept responsibility in directing the future of these infrastructure developments.

The tools, techniques and machines used in service development and delivery vary significantly across different service sectors. The concept of public service and its subset of civil service makes a clear distinction from services provided by the private sector [9]. Techniques and tools used in policing have little in common with those used in health services or educational services. Each sector must examine how the selection of tools and techniques address ecological and economic sustainability and how the utilization impacts community engagement and empowerment. This is a complex but necessary process that service workers must take on in collaboration with the communities they provide services. The selection of critical technologies and techniques can lead to increasing community control over service development and service provision. This is a goal of the appropriate technology movement.

Water and Sanitation

Safe water is fundamental to the survival and well-being of humanity. It must be seen as essential to prioritize the delivery of safe water globally. Currently, hundreds of millions do not have regular access to safe drinking water. Access to safe sanitation systems remains a problem across underdeveloped countries, particularly in Africa and Asia. Even more than twenty years after the

Table 2: Summary of Stakeholders role in Basic Services: Water & Sanitation

Stakeholder Category	Recommended actions
Education	K-12 education should address the importance of clean water, the fragility of water systems, university research and curriculum should address development of technologies that improve access to clean water and sanitation
Government	Safe drinking water and sanitation facilities to all must be an enforced policy. Subsidies must be provided for appropriate technologies in this area.
International Agencies	Promote national control of water and prevent cross border water theft. Regulate use of river water for benefit of masses across national boundaries. Provide international support for water and sanitation in poorest countries.
NGO/NPO	Provide water security projects at community level that employ appropriate technologies in water purification, irrigation and sanitation systems. Support school and local government projects that address water & sanitation.
Private Sector	Privately owned and public stock companies and corporations have an obligation to assure water and sanitation systems are developed and maintained in a manner that reduces inequality and empowers local communities. This is a priority over company profits and should be seen as a corporate social responsibility. Work closely with local communities to assure employee engagement with communities and company commitment to water security promoting ecological balance and appropriate technologies while reducing CO_2 emissions and mitigating climate change.
Citizen Activist	Community organizers must see control of their water resources as an essential right and work toward sustainability. Activists must monitor the water utilization by the private sector and make sure it is in line with the community's best interest. Also engagement with NGO/NPO, universities and government must technology transfer and appropriate technologies that assure water security

official end of apartheid in South Africa, rural communities still rely on unsanitary and unsafe pit toilets. Innovation in improving water and sanitation must be on the agenda of all stakeholders. Improving the monitoring of sewer system is one recent innovation[1]. Not enough research is devoted to developing innovations in water and sanitation access. This speaks to the inadequate attention and resources devoted to providing the masses with safe water and sanitation. It is known that water is a critical resource in the agriculture sector. Irrigation systems are the major contributing factor to why food production is higher in certain areas. Africa has more unused arable land, than other regions, that could be a major source of food production if proper irrigation was developed. Water is also an essential resource in mining operations, construction, and many manufacturing processes. In many regions and countries, water is limited and sometimes scarce. There are tradeoffs that must be considered how and when to use these water supplies. This is such an important concern that we have devoted a chapter to water and sanitation services and associated appropriate technology.

Energy

For centuries energy has been necessary to provide heated shelter and cooked food. Both are essential for the survival in so many communities. The technologies associated with providing energy for heating and cooking have advanced over the years. These technologies require energy. Gas, oil, coal, electricity and solar have been sources of energy for both heating homes and cooking food. Advances in the understanding of electricity led to the telegraph in 1844, followed later in the century with the invention of the telephone and light bulb. Delivering these and other electronic devices to a broad population required the development of electric power generation stations and power distribution networks [2].

National electric grids are the source of energy across the developed and developing world. In most underdeveloped countries access to these grids is limited to the most privileged citizens. Extending access to national electric grids must be a priority

Table 3: Summary of Stakeholders role in Basic Services: Energy

Stakeholder Category	Recommended actions
Education	K-12 education should provide fundamental understanding of energy systems, alternative energy systems research is key at university level
Government	Policies and tax incentives should support alternative energy at family, community, and state level. Decentralized and smart grid approach to large scale energy provisions. System approach to energy options must strengthen community engagement and control
International Agencies	Promote policies to curtail use of fossil fuel and nuclear power plants. Subsidize alternative energy projects in poor countries. Work closely with universities to promote national and international alternative energy projects.
NGO/NPO	Focus on alternative energy projects at community level with emphasis on underdeveloped countries and poor communities. Draw on university research and collaborations to advance appropriate technology in the energy sector.
Private Sector	Privately owned and public stock companies and corporations should develop non-fossil fuel energy systems not only for their operation but also to support the needs of local communities. This is a priority over company profits and should be seen as a corporate social responsibility. Work closely with universities to assure employee engagement and technology transfer that promotes innovation and research on energy systems.
Citizen Activist	Community organizers work for energy security and equality at affordable rates that reduces reliance on fossil fuel, promotes sustainability and CO_2 reduction.

in empowering the disadvantaged communities. One approach to extending electric grids is to develop local and decentralized grids. How the electricity is produced is a concern that must be given national and international priority. Electricity produced by coal-fired power stations is responsible for significant CO_2 generation. Extending these types of systems is not the proper solution to energy shortages. The cost of alternative low-pollution energy generation has reduced significantly over the past decade. Yet, the world remains slow in shifting away from fossil fuel power stations. Stakeholders must force policy changes to address this inertia than has the world stuck in a high CO_2 generation mode.

Housing

Cave paintings are some of the oldest representations of human culture. Caves were the earliest form of housing. Shelter from the elements of nature has been central to the survival of humankind. Housing for millenniums have been the responsibility of individuals, families, and small communities. Only in recent years has the idea of 'public housing' had any meaning. Even in 2022, it means little in much of the world. In high income countries, housing is the single most significant expense for most citizens. As countries advance technologically and increase their productivity, housing becomes more expensive and more elaborate. Housing has always been a clear sign of the inequality of wealth. King's castles to billionaire's mansion represent one end of the spectrum. Shanty towns and homeless sleeping in boxes on the street are the other end of housing reality. Various efforts by governments to provide public housing have met with mixed success. Cities to national governments have funded housing project to address the needs of workers, the poor and the homeless. Appropriate technology practitioners have explored low-cost housing, low energy housing and quick-build housing to address the needs of less privileged communities. Urban growth in both rich and poor countries require careful attention to planned housing. Homes and real estate are a significant source of wealth for upper- and middle-income families. If housing inequality is to be addressed all stakeholders must commit to policy changes and resource reallocation.

Table 4: Summary of Stakeholders role in Housing as a service

Stakeholder Category	Recommended actions
Education	Education systems at all levels should address training in building and maintaining homes. Appropriate building materials, mixed income housing, and appropriate architecture should be addressed in university instruction and research. Student projects and university projects should connect with disadvantaged communities to address housing inadequacies.
Government	Government policies and funding must support universal housing, appropriate architecture, and smart city housing. Special grants and tax incentives should support energy efficient and green constructed housing
International Agencies	Establish policies and practices that assure shelter for refugees, poor, women, and children. Address the needs of homeless and shanty dwellers with international projects that employ technology transfer and appropriate technologies
NGO/NPO	Focus on community level projects that employ appropriate technologies to replace shanty housing and provide quality shelter for the homeless. Special projects should address migrant communities, refugees and homeless, with special attention to women and children
Private Sector	Privately owned and public stock companies and corporations should develop green housing for employees and local communities as a commitment to corporate social responsibility. Work closely with civil society to promoted mixed income housing that reduces inequality and promotes sustainability and a circular economy
Citizen Activist	Activists must work with all other stakeholders to provide community-controlled housing that is sustainable and promotes circular economy and appropriate technologies. Assuring shelter for women, children and homeless should be given particular attention as a means of reducing inequality.

Transportation

Transportation as a service involves road, air and sea transport of both people and goods. Much attention is given to the need to shift from vehicles powered by diesel, and gasoline. The generation of carbon by these various forms of transport is a global concern. Electric vehicles powered by hydrogen fuel cells and batteries are clearly a positive alternative. However, in many cases the substitution of petroleum-based engines with electric motors does not reduce the carbon footprint. Such is the case in South Africa where most electricity is provided by coal-based power plants. An effective transformation of the transport industry must be holistic and address electricity generation, battery production, hydrogen production, as well as the carbon footprint of production of these alternative vehicles. Attention has been given to the development of rail industry for both transport of goods and people. Modern high-speed rail will connect major cities in countries and across national borders more efficiently than current systems. This is a core appropriate technology that contributes to reduced carbon emissions and improved travel safety.

Communications Cluster

Communications, entertainment, sports, recreation, and culture when considered as a service cluster is the fasting growing service sector globally. The linkage of these services goes back centuries. The public reading of newspapers was part of the culture in many countries in 19th century. It was both entertainment and source of shared news. The 2020 film *News of the World* stars Tom Hanks as news reader traveling town to town across West Texas in the USA in the 1870's. Today, the nexus of these services finds internationally televised sporting events embedded with advertisements from multinational corporations accessibly through various internet applications. This cluster of services oversee trillions of dollars of revenue globally. Little of this wealth trickles down to the local communities in the global south. Telegraph technology gave way to telephones, radio advanced from AM to FM and gave way to TV. TV was transformed by cable. The rise of the internet changed everything. During these advances, there was competition and the

Table 5: Summary of Stakeholders role in Transportation Services

Stakeholder Category	Recommended actions
Education	A start is addressing transport to schools to assure it employs appropriate technologies such as use of electric vehicles and low carbon public transport. Universities can focus on alternative energy vehicles, optimizing urban transport, smart city planning, and alternative approaches to long distance transport of goods and people.
Government	State owned transportation must focus on needs of local communities, poorer populations, and disabled. Policies must be established that support transport using alternative energy. City governments must commit to engraining smart transport into a Smart city agenda.
International Agencies	Policies that address improving international transport of goods, starting with restricted use of sanctions, to improve access to essentials to less developed countries. Reducing transport cost and CO_2 generation through shift to rail transport and employing appropriate technologies to reduce fuel use and CO_2 generation requires international cooperation.
NGO/NPO	Promoting innovation in transport at local, regional and cross-country level through use of appropriate technologies. Link with universities and community organizations on innovation and research for commuter transport in car, bus and rail.
Private Sector	Privately owned and public stock companies and corporations should take a systems approach to reorienting the transport mechanisms of their supply chain to focus on renewable energy transport and waste reduction. Work with communities to provide employees transportation options reduce CO_2 and inequalities, while promoting a circular economy
Citizen Activist	Community planned local transport can best engage appropriate technologies and systematically address inequality, sustainability and environmental issues. Activist must assure that transport plans in the context of smart city development and national infrastructure upgrades are consistent with the needs of the most disadvantaged and promote the circular economy.

rise of monopolies. Attempts by governments to counter monopoly saw only momentary success [7].

Table 6: Summary of Stakeholders role in Communication Services Cluster

Stakeholder Category	**Recommended actions**
Education	All levels of education must promote awareness of impact of recent and future information technology innovations on distribution of information and knowledge. Awareness of role of social media must lead to teaching and researching communication approaches that empower communities.
Government	Policy must break up multi-national monopolies in this cluster. Laws must be put in place to monitor practices and growth of MNCs in this cluster. The balance between freedom of speech and security & safety requires monitoring and policies.
International Agencies	Regulations on security, privacy and taxation of international communication systems should promote benefiting poorer communities and serve to reduce the digital gap in countries and between countries. Indigenous culture and rights must be respected and promoted in expanding media services. This requires use of appropriate technologies to increase empowerment of indigenous people in the communications space.
NGO/NPO	Partner with universities, community-based start-ups, and indigenous peoples' organizations to engage in new innovative communication practices that expand engagement and development of local communities, particular indigenous people.
Private Sector	Privately owned and public stock companies and corporations significant role in shaping media messages must be refocused toward promoting appropriate technologies to empower communities and eliminating all forms of inequality.
Citizen Activist	Activists must monitor the communication cluster to assure the balance between privacy, security, free speech and open access promotes democratic engagement and reduction in inequality.

Much of the media shared on the internet is provided by users who receive no renumeration. These users serve as unpaid workers in this service cluster. The paid workers in this cluster must find common ground with the unpaid user/workers. Communities across the world should study this service cluster to better understand its role in shaping the thinking and ideological perspectives of communities, organizations, and nations. The use of open-source technology and software is one example of appropriate technology to enhance the power of local communities. Media must be produced that promotes a factual representation of the current and historical class contradictions and wealth inequality. This will lead rectification of the situation and empowerment of communities. The most conscious progressive and revolutionary media workers must focus on grassroots organizing of this collective service sector. They could engage in local internet-based radio/TV that instills community empowerment. This would provide news, entertainment, and educational services. These outlets should be community controlled and seek to expand and link to similar outlets. For centuries the postal service has been a critical form of communication. Postal services have developed an extensive infrastructure of buildings, transport, logistics and people. Advances in information and communication technologies are transforming the postal sector. One perspective of the future role of postal structures is to assist in the transformation of urban settings to smart cities [6].

Education as a Service

Basic education is considered a human right and most nations attempt to provide universal education through the secondary level. State funded schooling for ten to twelve years is common. Access to quality education is still a major problem. While university education is expanding, it is still limited in most underdeveloped countries. It is suggested that universities can promote and improve their offerings by applying a service outlook [6].

Education is viewed by most developing and underdeveloped countries as the key service to promote national development, create jobs and reduce poverty. In most instances education has not been the panacea promised. It is essential to improve the assess to education and quality of education to make it a process to empower people and their communities. It is important to note

Table 7: Summary of Stakeholders role in Education Services

Stakeholder Category	**Recommended actions**
Education	The decolonization of education should shift the focus from 'conquest mentality' in all fields to a focus on empowering communities and safeguarding the planet's ecological balance
Government	Support of free education is important as well as forgiving student educational debts. Policies must be set that support appropriate technology, decolonization of education and community engagement.
International Agencies	Decolonization of education requires international support and policies that promote the role and culture of indigenous communities. International collaboration with university systems in former colonial countries where agencies serve to empower their communities, promote indigenous knowledge systems and apply appropriate technology
NGO/NPO	Collaboration at both K-12 and university level institutions at local and national levels should support innovative curriculum module content and instructional approaches, as well as promoting appropriate research at university level that deals with community needs.
Private Sector	Privately owned and public stock companies and corporations commitment to education, both K-12 and tertiary, must promote humanity-centered research and curricula, appropriate technologies, decolonization and elimination of inequality.
Citizen Activist	Community activists see community control of education as a priority. Working with both K-12 and university administration, workers and students is key to promoting curricula and research that is human-centered and engages appropriate technologies to assure sustainability and a circular economy.

that education is both a critical service and a critical stakeholder category. As a critical service education is addressed at all levels by the various stakeholder groups. Educators as a stakeholder group are engaged with appropriate technologies across the span of technology and service areas.

Table 8: Summary of Stakeholders role in Health Services

Stakeholder Category	Recommended actions
Education	Indigenous and endogenous medicine should be addressed at all levels. University research on telemedicine, expert systems linked to local health and medical expertise and resources is essential
Government	There is an obligation to provide free and low cost healthcare and give priority to support of preventive healthcare. Indigenous and holistic medical practice must be given proper recognition.
International Agencies	Addressing child mortality, basic healthcare in poor communities and the eradication of diseases such as malaria should involve agencies responding the needs of the poorest communities and countries. This can most effectively be achieved promoting technology transfer and cost effective appropriate technologies.
NGO/NPO	Work closely with community-based organizations to address most immediate health issues and engage university organizations and research to assist with human resources and appropriate technologies.
Private Sector	Privately owned and public stock companies and corporations must assure the health needs of all employees and assure that their workplace poses no health or safety hazards to employees or the community. There must be a commitment to provide reparations for any historical damages to the health of workers, community, or the environment.
Citizen Activist	Community activists can assert local control over local health provision to reduce inequality in healthcare access. Promoting healthy lifestyles, endogenous healthcare approaches while promoting technology transfer with other stakeholders is key to improving healthcare access for the most disadvantaged.

Health Services

Much like education, health services are provided by both the state and private sector and basic health service is considered a human right by international organizations like the United Nations and African Union. The weakness of health services in Africa and other less developed areas in the world results in a weakened labor force, higher child mortality and shorter life spans. It is necessary to change the incentives to politicians and service providers, to reshape the health service sector [7]. All stakeholders must challenge the current paradigms that organize public and private sector health services. The use of appropriate technologies and appropriate techniques provides the path to make health services effective in empowering people. How can recent development like telemedicine, robotics, expert systems, and other AI techniques be used to extend access to quality heath services? These and other technologies and techniques must lead to decentralized access to care, more accordable healthcare, as well as community control of these services.

Conclusions

As the significance of the service sector increases with urbanization and technology advances it becomes essential that the various stakeholders are held accountable for developing and enforcing service delivery policies that empower citizens and their communities. These practices must be equitable, environmentally friendly, and sustainable. Tables 2-8 highlight key aspects of the different stakeholder responsibilities relating to various services. These recommended actions are just a sampling of what stakeholder groups can do to make a difference.

This manifesto devotes an extensive chapter to services related to water delivery. This reflects the significance of water in maintaining healthy communities and addressing global inequalities. Water related appropriate technology can address the delivery of clean drinking water, the provision of sanitation services, and water needed for agriculture, mining, manufacturing, and transportation. The monitoring and balancing of the global water cycle are essential in combating climate change and environmental degradation. Policies and technologies to address both are critical to guaranteeing a global future where humanism flourishes.

The significance of the service sector and its varied strata require further research on how all stakeholder groups can engage each stratum. A closer examination must consider how certain service sectors will grow, some decline and others be totally transformed. The possible scenarios must be considered with a mindset to empower communities, promote sustainable appropriate technology, and eliminate all forms of inequality.

Bibliography

[1] I. Gough, "Universal Basic Services: A Theoretical and Moral Framework," *The Political Quarterly,* vol. 90, no. 3, pp. 534-542, 2019.

[2] G. Martin, "DefenceWeb," 20 April 2021. [Online]. Available: https://www.defenceweb.co.za/featured/900-service-delivery-protests-in-south-africa-over-six-months/. [Accessed 14 January 2022].

[3] J. 2. 2. Aaron O'Neill, "Statista," 27 July 2021. [Online]. Available: https://www.statista.com/statistics/578944/employment-by-economic-sector-in-south-africa/. [Accessed 19 December 2021].

[4] "U.S. BUREAU OF LABOR STATISTICS," 8 September 2021. [Online]. Available: https://www.bls.gov/emp/tables/employment-by-major-industry-sector.htm. [Accessed 19 December 2021].

[5] A. O'Neill, "Statista," 23 July 2021. [Online]. Available: https://www.statista.com/statistics/270382/distribution-of-the-workforce-across-economic-sectors-in-the-united-kingdom/. [Accessed 19 December 2021].

[6] M. Szmigiera, "Statista," 18 March 2021. [Online]. Available: https://www.statista.com/statistics/962868/global-employment-region-sector/. [Accessed 19 December 2021].

[7] A. K. Shittu, "Public Service and Service Delivery," in *Global Encyclopedia of Public Administration, Public Policy, and Governance,* Springer, 2020.

[8] J. Maluleke, "Improving Services by Sewer Asset Condition Assessment," *Journal for Creativity, Innovation and Social Entrepreneurship,* vol. 5, no. 1, pp. 48-64, 2021.

[9] C. R. Roach, Simply Electrifying The Technology that Transformed the World, from Benjamin Franklin to elon Musk, Dallas: BenBella Books, 2017.

[10] T. Wu, The Master Switch The Rise and Fall of Information Empires, New York: Alfred A Knopf, 2010.

[11] J. Trimble, "The Post Office in the Smart City," in *Exploring New Frontiers: Navigating the Evolving Digital Landscape*, Oxfordshire, Libri Publishing, 2021, pp. 157-162.

[12] J. F. Irene C L Ng, "EDUCATION AS SERVICE: THE UNDERSTANDING OF UNIVERSITY EXPERIENCE THROUGH THE SERVICE LOGIC," *Journal of Marketing of Higher Education,* vol. 2009, no. May, 2009.

[13] P. B. a. H. Rogers, "SPOTLIGHT 8 Service delivery: Education and health," in *World Development Report 2017*, WDR 2017 team, 2017, pp. 189-193.

CHAPTER 5

Appropriate Technologies for Water and Sanitation[1]

John Tharakan

This chapter focuses on appropriate technology as it pertains to water and sanitation. Potable water availability and sanitary treatment and disposal of wastes are two critical prerequisites for the improvement of public health, and for the development and maintenance of healthy, viable and sustainable communities. In this chapter, we review appropriate technologies that are in development and implementation for water sourcing, collection, treatment, storage, conservation and management. The review is conducted through the philosophical and ethical lens of appropriate technology, in the context of sustainable development and community wellbeing. This chapter highlights those water technologies that would make this possible. In addition, appropriate technologies for sanitation are assessed and evaluated. Conventional methods of waste disposal including land filling and incineration, while offering short-term solutions to the problem of increasing waste generation, have severe adverse environmental impacts. More appropriate waste management technologies including biologically based processes that harness the potential of biological agents such as plants, microbes and earthworms, to treat contaminated effluents from industry as well as to remediate and decontaminate hazardous and contaminated sites, are considered. These appropriate technologies for sanitation address the waste problem by a paradigm shift that recast wastes as resources and translates the discussion from one of "how to dispose of wastes?" to one of "what technologies will enable utilization of these wastes as resources to add value". Measures of public health will only show improvement with the development and implementation of diverse but pertinent and appropriate technologies for water treatment and availability, maintenance of adequate water supplies, and sanitary and sustainable treatment of wastes. In this chapter, we review and evaluate various "appropriate" water technologies, highlighting some of the best available technologies and practices, discussing their application in, and pertinence to, sustainable development, improving the quality of life and contributing to enhanced wellbeing.

Introduction

More than two decades ago this past summer, the United Nations (UN) General Assembly voted overwhelmingly to endorse "...the right to safe and clean drinking water and sanitation as a human right that is essential for the full enjoyment of life and all human rights"[2]. One hundred and twenty-two countries supported the resolution and not one single country opposed it. Forty countries, including some of the world's wealthiest democracies such as the United States, the United Kingdom, several European countries as well as Australia and New Zealand abstained; several of these countries instead pushed for a watered-down UN Declaration that would declare "access" to water a human right. To the UN's credit, the Bolivian United Nations Ambassador, who put forth the resolution, resisted, arguing that simply arguing for "access" would not ensure availability, especially in the context of water as a commodity that would need to be purchased even if access is provided. An especially current instance of the struggle for the right to water is exemplified by what happened in Ghana, where citizens fought and succeeded in the removal of prepaid water meters as part of an ongoing fight for the human right to water (Shang-Quartey, 2017).

The Human Right to Water and Sanitation laid out some very disturbing facts about the water and sanitation situation on the planet as it was over a decade ago, with over a billion people – one out of every six humans at the time – "...lack access to safe drinking water; over 2.6 billion do not have access to basic sanitation, and approximately 1.5 million children under 5 years of age die and 443 million school days are lost each year from water and sanitation related diseases." The right to water has been articulated and codified before on numerous occasions by various international bodies (see for example, The World Health Organization[3]) and this history is also outlined in the UN Resolution on Water and Sanitation that was just approved by the UN General Body.

Decades of experience as documented by development experts, and understood through common sense by lay people, have shown that availability of and access to clean water and engineered treated sanitation are perhaps two of the most critical requirements for establishment of healthy, viable and sustainable communities and their well-being. Without clean potable water, there is no support for life. And with no sanitation, communities are being sickened, poisoned and overwhelmed by their own wastes. In the developed world, these are taken for granted. One need only turn on a tap to get clean potable water or push a button or press a lever to sanitarily wash away one's wastes. Much of the developing world, especially poor urban and rural communities, lack these two basic necessities of a healthy life. Images from the developing world of people swarming around a water tanker or crowded around a single water pump all with waterpots in hand in a dry, parched and barren landscape are legion; so are images of open sewers and waste pits and piles in the mega-cities, towns and villages of the global south, where untreated effluents flow over pavements and streets and piles of waste smoke and smolder in hazy, dirty conditions, while rag pickers walk over these piles in the third world's most widely used configuration of a resource recovery systems (Pushkaran, 2013).

Clean, potable water availability and environmentally benign sanitation are critical infrastructural needs for the development and maintenance of public health. Numerous organizations and individuals, including this author (Tharakan, 2006) have argued for the establishment and codification of the basic human right to water – with the basic argument and rationale that without water there can be no life. Now, as outlined above, the UN has declared it to be a basic human right, along with sanitation. Hence, despite the abstentions of developed countries more interested in promoting the privatization and commercialization of water focused specifically on supporting and enhancing profits for the large multinational water companies based in the Global North, we have a consensus from the global South that access to clean water should be a basic human right.

Enshrining a human right to water in a UN declaration is indeed as it should be. The corollary to this codification of the right to water should be the investment of resources by governments and non-governmental organizations (NGO's). This suggests the need for widespread development and dispersion and transfer of appropriate technologies to ensure that communities have access to clean water and sustainable sanitation.

Appropriate Technology

The widespread use of the term "appropriate technologies" requires a discussion and articulation of what exactly it means for a technology to be deemed "appropriate". Indeed, appropriate technology, or AT for the rest of this paper, has always been difficult to define and its development and implementation have also been a source of debate for some time [Darrow and Saxenian, 1986]. Nevertheless, there is general agreement that AT empower local communities, utilize local resources, and be affordable. AT has to be grounded in the community, utilize indigenous knowledge, be comprehensible, controllable and maintainable through local skill development and community capacity building. Any appropriate technology should also be adaptable, must include local communities in the innovation and implementation stages, while minimizing adverse environmental impacts [Darrow and Saxenian, 1986]. Earlier papers [Tharakan, 2004; Tharakan, 2010] have provided broad over views of appropriate technologies available for water collection, treatment and storage in the contexts of public health and more broadly across the social space.

It is increasingly recognized that water will be the "gold" of the 21st century. The World Water Forum's (WWF, 2003) Ministerial Declaration[4] stressed access to clean water as the driving force for sustainable development and the eradication of poverty. In 2004, the People's World Water Forum (PWWF) convened in New Delhi, India with more emphasis on the need to recognize access to potable water as a human right, while articulating the strong opposition to privatization of water resources from grassroots communities across the world. The PWWF noted the critical importance for the local community retaining control over water resources for sustainable development, social empowerment and community wellbeing. Subsequent People's World Water Forums have only underscored this right and argued for universal access and availability of clean drinking water across the planet. Shiva [2002], in *Water Wars: Privatization, Pollution and Profit*, has critically examined water rights where she has argued that the "market paradigm" for water use and distribution is flawed, numerous examples of which now populate the development history landscape, where the take over of public water utilities by private capital has, more often than not, resulted generally in reduction of services, increases in water rates and increasing issues of the maintenance of water quality and standards. Shiva's summarization of this history demonstrates that the corporate control over water utilities and resources results in the erosion not only of water resources but also of democratic structures that have historically governed community water use.

Now, with the UN declaring the human right to water, there is new impetus to push forward with the development and dissemination of appropriate technologies for water and sanitation. This chapter focuses on reviewing and evaluating appropriate technologies for water and sanitation. There have been numerous collections of works on water policy, technology and development, some coming from the Center for Science and Environment (CSE) based in New Delhi, India [Agarwal et al, 2001]. CSE developed and implemented groundbreaking and creative solutions to water resource management that garnered them the Stockholm World Water Prize in August 2005[5]. Given the importance of water conservation and recovery, water harvesting is the first technology that must be dispersed and diffused through out the world.

Water Harvesting

Water precipitating out of the sky in the form of rain, snow, sleet, hail or other precipitation percolates through the ground to replenish groundwater and feed subsurface aquifers and streams. The precipitation runs off impervious surfaces until it reaches a surface water body or pervious soil surface where it can percolate into the ground water. Or it can evaporate directly or through transpiration back into the atmosphere. Although the nature of the hydrologic cycle makes it difficult to mark a beginning or end to waters cyclic journey through the environment, rain can be considered a primary source of water. In this context, secondary sources of water include rivers, lakes and groundwater, all of which get recharged from the primary water source through time. Most development experts and technocrats tend to focus more on the secondary water sources as the major input streams to water collection, treatment and storage systems. This ignores the primary water source – which feeds all the secondary water sources – and hence such thinking undervalues or devalues the importance of the primary water source and of incorporating primary water sources into the design and development of water resource systems. Through rainwater harvesting, the value and importance of rainwater is reestablished and incorporating water harvesting into the spectrum of potential water sourcing technologies brings the development and harnessing of this precious resource into the realm of sustainability [Keller, 1984].

The principles underlying rainwater harvesting and the calculations that enter into the determination of the design are straightforward. If one knows the amount of rainfall that an area receives, multiplying this by the efficiency with which the rainfall can be collected (harvested) will provide the potential amount of water that can be harvested. Usually, the amount of rainfall multiplied by the area of 'catchment' will provide the volume of water that can be collected. Following the discussion in A Water Harvesting Manual [Kellar, 1984] as an example, a rooftop with an area of 100 sq.m receiving 2200 mm of rain in a year could potentially provide 220 cu.m. (220,000 liters) of water. If the water harvesting system design permitted 60% efficiency in collection, then at least 132,000 liters would be available.

In its simplest form, the basic elements of a simple rainwater harvesting system are shown in Figure 1 [CSE, 2003]. The catchment area is established and then a conduit or pipe is connected to this area enabling the water to be sent to a storage facility and to a recharge facility. The storage facility provides immediate water for ready use and can be below ground or above ground, while the recharge facility provides a mechanism by which longer-term water storage can be recharged for later withdrawal.

Figure 1: Basic elements of a rainwater harvesting system. From [CSE, 2003]

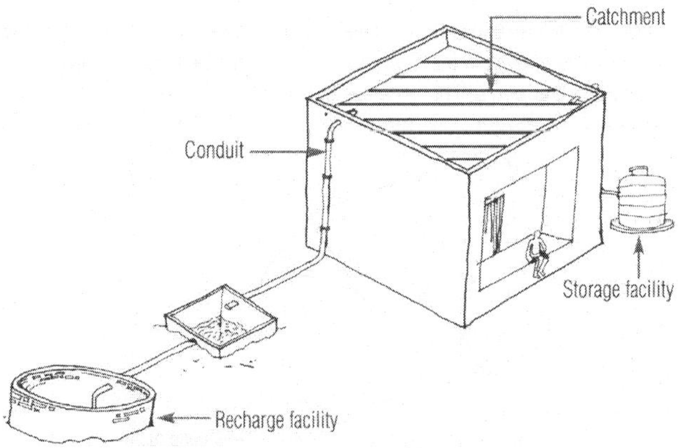

Since falling rainwater will bring with it any dust, particles and debris that was in the atmosphere or on the catchment area, it is beneficial to insert a filtration mechanism at the head of the conduit collecting from the catchment area to trap all such materials. This level of filtration is often sufficient for many water uses. Additional filtration will be required for potable water, and slow sand filtration is often adequate for this end-use [Van Dijyk and Oomen, 1978]. Slow sand filtration will clean water supply sufficiently to make a significant improvement in public health. For complete elimination of pathogenic organisms and to ensure that public health is maintained through the elimination of unclean water as a disease transmission vehicle, disinfection of the water will be required. This may be accomplished through various additional point-of-use technologies such as boiling, chemical disinfection or filtration. Boiling is perhaps the most effective in sterilizing water but energy requirements are high and will add to the cost of the water to the public user. Requiring the public consumer to boil their water prior to consumption will carry the risk of many failing to do so and thus raising the risk to maintenance of public health. Chemical disinfection is often the quickest but requires the addition of a disinfecting agent such as chlorine, which kills all types of bacteria and makes water safe for drinking. However, chlorination of water is known to generate harmful byproducts and this method, although easiest and cheaper then boiling, may have long-term adverse consequences for public health, and appropriate and sustainable alternative technologies need to be researched, developed and implemented with scale-up potential always in mind.

Perhaps the simplest and least expensive method is through in-line filtration devices built into the water harvesting, collection and storage system design. A simple charcoal filter can be constructed of sand and charcoal layered one over the other sandwiched between two coarse-pebble or gravel layers that facilitates percolation of water and prevents clogging of the filter. The sand efficiently removes particulates and the charcoal acts as an adsorbent to bind microbial contaminants, other colloidal and suspended contaminants and also serves to remove organics and metals. An example of a simple low-cost filtration media configuration is shown in Figure 2. This filtration set up has been shown to remove pathogenic bacteria as well as other microbes such as parasites and amoebas, the causative agents in dysentery and diarrhea.

There are other simple methods of disinfecting water, such as through distillation. A recently developed low-cost, low-maintenance solar disinfection unit has demonstrated quite a bit of promise as an appropriate, low cost technology for the production of potable, disinfected water. This unit eradicated over 99.99% of bacteria in water samples and was able to provide six liters of pure drinking water on a daily basis [Jagdeesh, 2006]. Such decentralized and stand-alone single-user appropriate technologies need to be expanded and developed further, always with an eye to scaling and dissemination of the technology. These AT's can also be incorporated into water resource development schemes to benefit local communities and people at the grass roots.

Figure 2: Basic elements of a sand-charcoal filter configuration. From [CSE, 2003]. The size of the filtration set-up can be increased with ease and material costs are very low.

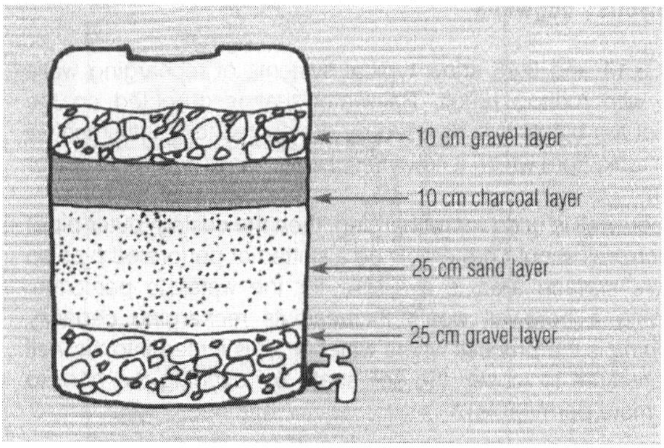

The potential for problems with contamination of the stored water are real but can be avoided with the implementation of suitable measures to minimize risk and prevent contamination. It is important that storage tanks are cleaned prior to the start of the rainy season and the catchment area be swept and cleaned as well. Conduits for conveyance of the harvested water to the storage receptacles should be kept clean and unclogged. The in-line filtration apparatus must also be cleaned and maintained so that the water can freely percolate through the media. This might require periodic cleaning out of the filter. In terms of end-use, it would be necessary to change the charcoal and sand media in the sand-charcoal filter on a periodic basis. When the filter media are changed, it is important that the new filter media be flushed completely prior to the water being used as potable water.

The design of the water harvesting system and the configuration of the filtration/disinfection set-up will depend on the context within which the water collection system is being developed and implemented. Designs and systems for rural areas will be very different from those that would be implemented in urban settings. Further, rural riverine settings will require far different designs and configurations than rural mountainous or arid regions. Despite this, the basic components of these systems – a catchment area, conduits to channel the harvested water, means for filtration and disinfection, and storage reservoirs – will be the same. As in the development and implementation of any appropriate technology, the specific system to be established and the specific design to be implemented and constructed will necessarily be highly dependent on the local situation. The configuration that is finally settled on within a particular context must be thoroughly examined and tested through actual use. Amounts of water that are harvested, water quality and the time required for collection must be recorded and these results need to be evaluated after a given period of use. Depending on the situation and the local context, there may be a need for redesign and reconfiguration. This is part of the process of technology development and implementation and must not be neglected so as to optimize the final design that is implemented on a larger scale within a given region and context.

As with the process of any development technology that is being designed and implemented, serious and critical consideration must be given to how well the local community's needs are being met and what the benefits and costs of the technology implementation are. Care must be taken that the community feedback is taken into account and that the community is itself engaged in the entire process. Successful development and implementation of appropriate technologies will only result if the local community that the technology will serve is engaged in the process from the outset. This would necessarily include community training and knowledge and technology transfer so that community input forms a substantive and integral part of the design, development and implementation process.

Sustainable water management technologies is being broadly researched and supported, such as the Institute for Sustainable Technologies in Germany (Atasoy et al, 2008) which has looked at water re-use in small municipalities and investigated the use of RWH in diverse socio-ecological and politically charged settings, such as the occupied Palestinian Territories.

Appropriate Technologies for Sanitation

Determining what to do with our wastes is another critical issue facing developing communities and emerging economies. Conventional methods of waste treatment and management include, and are usually limited to, incineration or land filling for solid wastes, and discharge into sewage treatment systems that utilize conventional and well established wastewater treatment technologies and management systems, for liquid effluents and wastes (Tharakan and Ismail, 2007).

For liquid wastes, conventional sewage treatment systems are capital intensive and require large infrastructural investments that are beyond the reach of most developing communities. More appropriate wastewater treatment technologies need to be developed and implemented, such as the DEWAT[6] systems installed in Pondicherry in south India, referring to decentralized wastewater treatments system. DEWATS applications are based on the principle of low-maintenance since most important parts of the system work without technical external energy inputs and cannot be switched off intentionally – these systems epitomize the characteristics of an appropriate sanitation technology as they provide a state-of-the-art-technology at affordable prices using local materials. DEWATs includes primary treatment with sedimentation and floatation followed by secondary anaerobic treatment in fixed-bed reactors (either baffled upstream reactors or anaerobic filters) and finished with tertiary aerobic treatment in sub-surface flow filters or in polishing ponds. DEWATS treated water meets requirements stipulated in environmental laws and regulations.

Wastewater can be treated to different specifications, depending on the situational context, including geographic, hydrologic, and socio-economic considerations and assessments. Part of improving the sustainability of water use is to be able to re-use waste water by minimal treatment, treating the waste water as an actual resource, such as demonstrated by Dharwish et al (2014), who have shown how in Qatar, one of the world's poorest countries in terms of fresh water resources, and one with the highest water consumption per capita in the world (>500 L/day/person), the high volumes of wastewater are treated using advanced technologies, and the quality of the treated wastewater (TWW) achieved allows for the water to meet standards for irrigation use, and they have shown, using scientific, economic and technical evidence, that TWW is a valuable and safe resource for crop irrigation that poses minimal risk to the soil, GW, and crops and is a key factor towards Qatar's food security.

In Africa, Amigun et al (2012), have rigorously demonstrated that wastes from rural Africa can be utilized for anaerobic degradation and generation of biogas for energy as well as the effluent for agricultural fertilizer application. Anaerobic digesters produce biogas from organic wastes and represent one of a number of village-scale technologies that offer the technical possibility of more decentralized approaches to development. In addition, this technology offers a very attractive route to utilize certain categories of biomass such as agricultural organic waste or manure in rural areas for partially meeting energy needs including heating and electricity. This sustainable and appropriate technology can therefore serve as a means to overcome energy poverty, which poses a constant barrier to social and economic development in developing countries.

For solid waste, utilizing biological methods of waste treatment and management provides environmental, cost and social benefits over and above conventional incineration and land filling technologies. With incineration, aside from the tremendous expense, one has to be concerned about the production of dioxins and other extremely hazardous compounds as well as the disposal of the toxic ash that is usually high in concentration of heavy metals and other hazardous contaminants. When the land filling option is chosen, concerns about leaching and groundwater contamination, methane eruptions and volatilization of contaminants emerge and become significant.

Choosing an Appropriate Technology

There are numerous appropriate technologies that exist for water collection, treatment, and conservation. Many of these have a long history, having been developed and used for centuries by people indigenous to particular regions [Kellar, 1984], and these have served communities extremely well. More appropriate and sustainable technologies have also been extensively described and catalogued [Schulz and Okun, 1984]. In the development and implementation of appropriate technologies, often the initial selection of the 'right' appropriate technology from a range of different choices is the key element in determining long-term successful implementation, adoption and operation of the chosen technology. There is a range of analytical heuristics that one can apply to the process of making the decision as to which technology would be best, which technology choice would in fact be the most sustainable. There has been considerable research done on technology choices in developing communities [Parr and Shaw, 2006], where the analysis proceeds through three separate stages.

In the first stage, the objectives are clarified and established. The question here is straightforward – what exactly is trying to be achieved? Some corollary questions follow, such as whether the goal is achievable and realistic and whether this is indeed the solution to the primary problem. In the case of water treatment for public health, the goal is straightforward. Simply put, it is that clean potable water be provided to the entire population and equitably across all communities.

The second stage of the technology selection process is analysis. In this stage, it is important to identify what the constraints on the proposed developments might be. Constraints here are not limited to physical constraints. Certainly, it is important that those be identified, including what are the actual water resources available, what is the land availability and any other physical constraints that would impact technology choice, development and implementation. However, just as important are a set of other constraints, including social, health, technological, economic, financial, institutional and environmental. These various constraints have been bundled into a concise acronym for analysis, called SHTEFIE [Parr and Shaw, 2006]. This heuristic algorithm for analysis and assessment was originally developed at Loughborough University as a tool for the evaluation of various development program alternatives. Once this analysis is completed and all the various elements that would impact the success of a particular technology choice have been evaluated and assessed, it is then possible to make a decision on technology selection.

The selection is the third and final stage of the analysis process and is termed the output. The output will clearly answer the question as to what particular technology is appropriate, given the problems and constraints of the particular context – social, health, technological, economic, financial, institutional and environmental – where the technology is sought to be implemented.

Conclusion

Access to clean potable water and adequate supplies of the same are critical to the development and maintenance of public health. Use of appropriate technologies to develop water resources and make clean water available to all is crucially important to the maintenance of public health. The design, development and implementation of specific water harvesting systems must take into account the context-specific situations and factor in community and infrastructural considerations as the water resource technologies are developed and put in place. Rigorous analysis and evaluation using the SHTEFIE process of the various inputs, situational variables and parameters must be conducted to reveal what is the most appropriate and optimum technology choice for water treatment in a given situation.

Given that the major stakeholder groups that are being addressed through the Appropriate Technology manifesto include:

1) Academia and educational institutions; 2) government agencies; 3) multilateral institutions and agencies; 4) non-governmental organizations (NGO's), non-profit organizations (NPO's), and community-based organizations (CBO's); 5) the private sector; and 6) engaged citizens, these stakeholder's responsibilities ensue.

For water security and equity, academia and educational institutions must focus resources on supporting faculty and students engaged in water research and support academic extension programs that leverage student engagement through project-based service learning that provides a pathway for student teams to work with communities in addressing water resource problems. Educational institutions must support innovation and incubation activities that provide the scientific and technological research and development of appropriate technologies for water, such as the Water Lab at the Indian Institute of Technology, Madras (Pradeep, 2021). In addition, knowledge transfer in terms of teaching and training, outreach and extension work in the communities must be part and parcel of any academic institutions engagement.

Government agencies must support the research, development as well as the implementation of appropriate water treatment and storage technologies across the diverse geographies that may be part of any national geographical region. These government agencies must partner constructively with multilateral institutions and agencies to leverage the needed financial and other support services to make the implementation of appropriate technologies a priority. At the same time, NGO's, NPO's and CBO's must all be engaged to ensure that the services and products that they seek to deliver or develop within communities utilize the most appropriate and sustainable technologies available, and that they engage in not only the deployment of the alternative technologies for water and sanitation but also the knowledge and technology transfer that is critical to undergird the sustainability of any of these technological deployments or interventions. AT must not suffer the fate of the 1000's of broken and unusable hand pumps installed by development aid in poor and under resourced communities.

Finally, the Private sector as a stakeholder needs to engage in corporate social responsibility beyond the green-washing statements often put out by the publicity departments of polluting corporations. The corporate social responsibility must be engaged in developing and deploying clean water technologies across the communities they profit from in direct and indirect ways as part of paying back and giving back, an ethical approach that must be inculcated into the private sector through education and outreach from academia as well as government agencies and institutions that often provide seed and ongoing funding for these corporate ventures.

Ultimately, the goal should be clean water and sanitation for all through supporting the communities and the citizen activists that reside in these communities and are engaged in improving the quality of life.

Refererences

Agarwal, A., S. narain and I. Khurana, Making Water Everybody's Business, Center for Science and Environment, New Delhi, India (2001)

Amigun, B., Parawira, W., K., J., O., A., & S., A. (2012). Anaerobic Biogas Generation for Rural Area Energy Provision in Africa. Biogas. https://doi.org/10.5772/32630

Atasoy, E, I. Bergmann, B. Regelsberger, M. Regelsberger (eds) Sustainable Water Management, Institute for Sustainable Technologies, Gleisdorf, Germany, January, 2008. [http://www.aee-intec.at, http://www.zero-m.org].

Brush, R.E., Well Construction, Peace Corps Appropriate Technology for Development Series, Washington, DC (1971).

Cairncross, S., and R. Feachem, Small Water Supplies, Ross Institute, London School of Hygiene and Tropical Medicine, London (1978).

Center for Science and Environment, A Water Harvesting Manual for Urban Areas: Case Studies from Delhi and Mumbai, p. 3, CSE, New Delhi, 2003.

Darrow and Saxenian, Appropriate Technology Sourcebook, Volunteers in Asia, Stanford, CA (1986).

Darwish, H, H. Abdulrahim, A.N. Mabrouk, A. Hassan, B. Shomar, "Reclaimed Wastewater for Agricultural Irrigation in Qatar," *Global J.Agric.Res.and Reviews,* 3(1) 106-120, 2015.

Jagadeesh, A, "Drinking Water for All", Center for Energy and Sustainable Resources, R.M.K.Engineering College, Kavaraipettai, India (2006); http://www.ewb-international.org/pdf/WaterForAllJagadeesh.pdf ,

Keller, K., Rainwater Harvesting for Domestic Water Supply in Developing Countries: A literature Survey, U.S. Agency for International Development, Arlington, VA (1984).

Koegel, R.G., Self-Help Wells, Food and Agricultural Organization, UN, New York (1978).

Pacey, Arnold, Hand-Pump Maintenance, ITDG, London, (1980).

Parr, J., and R. Shaw, "Choosing an Appropriate Technology," http://www.lboro.ac.uk/departments/cv/wedc/, Accessed June 2006.

Pradeep, T., "Exploring Opportunities with Advanced Materials" (2021) Indian Institute of Technology Madras, Pradeep Research Group. [Available at https://pradeepresearch.org/other-initiatives/iccw] International Center for Clean Water.

Pushkaran, JA, "No Room For Recycling – Give Waste Pickers Their Due," (2013) *Down to Earth,* New Delhi, India [Available at: https://www.downtoearth.org.in/blog/no-room-for-recycling-42476 - Accessed Feb 25, 2022].

Rybcynzski, W, Paper Heroes; Appropriate Technology: panacea or Pipedream, Penguin, USA (1991).

Shang-Quartey, L, "The Fall of Pre-paid Water Meters in Ghana," (2017) Grass Roots Alternative Development Center, Accra, Ghana [Available at: http://www.blueplanetproject.net/wordpress/wp-content/uploads/WaterMeters-Ghana-3.pdf; Accessed Feb25, 2022].

Shulz, C.R., and D.A. Okun, Surface Water Treatment for Communities in Developing Countries, John Wiley and Sons/IT Publications, London, 1984.

Siva, Vandana, Water Wars:Privatization, Pollution and Profit, South End Press, Boston, MA, USA 2002.

Tharakan, J., "Appropriate Technologies for Water Use and Conservation in Public Health," Proc.2nd Intl.Conf.Appropriate

Technology, July 2006, National University of Science and Technology, Bulawayo, Zimbabwe, pp 87 – 92 (2006).

Tharakan, J., "Appropriate Technology and Water Availability and Use: Impact on and Implications for Land Reform," Proc.1st Intl. Conf. Appropriate Technology, July 2004, pp. 97 – 104, National University of Science and Technology Press, Bulawayo, Zimbabwe (2004).

Tharakan, J, "Appropriate Technology for Water and Sanitation," in Dzidzeniyo et al (eds),*Proc. 4th Int'l Conference on Appropriate Technology*, Accra, Ghana, November 2010, *pp 5– 12.*

Tharakan, J. and SA Ismail (eds), Proceedings of Symposium on Biological Methods of Waste Treatment and Management in South India, The New College, Chennai, February 2007.

Van Dijk, J.C., and J.H.C.M.Oomen, Slow Sand Filtration for Community Water Supply in Developing Countries: A design and Construction Manual, Technical Paper No. 11, IRC Press, London (1978).

Endnotes

1 Portions of this chapter have been adapted from papers previously presented at the International Conferences on Appropriate Technology, 5th and 6th [http://www.appropriatetech.net]

2 For the draft UN resolution, please see: https://www.blueplanetproject.net/index.php/home/water-movements/the-right-to-water-is-the-right-to-life/

3 https://www.who.int/health-topics/water-sanitation-and-hygiene-wash

4 http://sdg.iisd.org/news/third-world-water-forum/

5 https://siwi.org/latest/stockholm-water-prize-the-supply-is-in-the-sky/

6 DEWATs systems are supported by the Bremen Overseas Research and Development Organization, https://www.borda.org/solutions/decentralised-sanitation-systems-2/

Chapter 6

Appropriate Technology, Manufacturing, and Transportation: A Conversation

Prof. K. Mpofu (KM) and Prof. J. Trimble (JT)

This conversation is to highlight the importance of industry, in particular the manufacturing and transportation sectors, in employing technology. Manufacturing has a rich history as being central to advancement of society. The first industrial revolution saw the shift from rural and farming employment to steam powered manufacturing. Since then the 2nd, 3rd and now 4th Industrial revolutions have advanced the manufacturing sector in uncountable ways.

An important subsector of manufacturing is the transport manufacturing sector. Given the colonial legacy of segregation of populations based on skin pigmentation this legacy has often left the indigenous people distant from the means of access to basic quality services such as work, education and healthcare. Thus, the production of safe, affordable and accessible modes of transportation is foundational in enhancing the quality of life of populations in the global south. South Africa is a classical case given the deliberate apartheid state endeavor to distant the black population from the central business district and the industrial vicinity. The provision of affordable transport is highly dependent on the manufacture of the modes of transport across the entire life cycle of travel.

JT: Highlight major advances in manufacturing development and how they have impacted communities and countries. When and how did certain advances empower people and their communities? We want to speak about this from continental perspective and global perspective. We are primarily concerned with focusing on technology changes, particularly transportation and manufacturing, and how these changes can be directed to empower Africans and the global south. We realise since technology is more pervasive in western countries that they have more control over the direction of technology advances. Therefore, we must be concerned with technology from a global perspective.

KM: Where I am seated, development associated with the 1st through the 4th Industrial revolution, which come with their particular frames of how technology advances, there has always been a means and ways of maneuvering the use of technology. An example is crossing of rivers in Africa. People have found ways of dealing with needed river crossing without necessarily relying on western technology or the most advanced technology. They employed indigenous approaches. Now with access to more advanced technology, communities have been creative in combining the old with the new. However, in many instances people are stuck on one side of the river unable to get across to get access to the services needed. In the Eastern Cape the communities have been contacted to work with my Research Chair to see what interventions can be used. One example of crossing now is the use of large metal Drums, used for storage. People get in the drum and cross with assistance of a good swimmer. This is risky and alternatives must be developed. This highlights the current reality in many rural areas and also highlights the challenges in what is needed in terms of appropriate technology. The significant technologies are introduction of Automotive Combustion Engine (ACE). This was a basic ingredient in modes of transportation for many years. Since then, we have seen advances with alternative energy sources powering vehicles from electricity to hydrogen. We see now the development of driverless vehicles. All these developments will have global impacts. The other realities we see in South Africa, given that SA is more developed than most of Africa, we have here some of the poorest of the poor. This means they experience the same problems we see in poorer countries in Africa regarding transportation not to mention the global South at large. The developments we make to assist the poor in SA can be applied across Africa and in the global South in general.

JT: How has manufacturing and transportation failed to address environmental concerns (industrial waste, inefficient use of power systems, etc)? One of the things we focus on in INAT is much of manufacturing and transportation fail to address environmental damage.

KM: Here in SA one of the main challenges, in the more sustainable and affordable mode of transport has been the train. However, with narrow gauge track construction the speed of trains is limited. This means to address high speed travel we need to look at track infrastructure in addition to train construction itself. The reality is from a manufacturing of trains it is important, but construction of the rail infrastructure system is pivotal. This is an important space for dealing with inequality, unemployment and job creation. The reduction of car transport that can be achieved with expansion of commuter rail systems is important. Not only in the service it will provide for millions of Africans, but in the reduction of GHG emissions locally, regional and the global South in perspective.

JT: this use of train transport, particularly electric train transport will be a very good contributor to the reduction in pollutants.

What suggestions should be given priority in reshaping manufacturing to become more environmental responsible beyond the rail transport sector?

KM: Also speaking to the modes of transport that speak to mass use, we should look at advances in the use of bicycles. While with large distances to work this may be less practical a combination of multiple modes would be ideal. However, with the design of smart cities and the practicality on smaller cities, the use of bikes can fit well. This works well with low skilled sectors. Selling this to other countries in the global south can be useful for TUT and an important entrepreneurial opportunity. Another project at TUT is the Solar car project that has worked at TUT for years. Recently a team drove an electric -solar car to the coast of Namibia in Swakopmund. I must highlight safety issues regarding use of bicycles.

I witnessed an incident where cyclist was hit by car and it was bad. This requires policies on the use of bicycles and infrastructure that support travel on bicycles. Both cyclists and automotive drivers but have clear guidelines and understand the importance of following these to maximize safety. Regarding aircraft, a small company called Sling are engaged in dealing with small craft that can be used for transport of products if the needed standardization processes are followed.

JT: That highlights other types of transport, but what about industries, like manufacturing of building materials and construction of building.

KM: When you think of the different ways of building construction, most of the building was done with local material. Now more construction relies on purchase of non-local materials. This requires transport contributing more to GHG generation. How can we deal more with local materials? One local company is using recycled plastic for construction material. This is important in using recycling to reduce GHG. Local companies need to do more of this. The balance between supply of recyclable materials and the need for transport manufacturing materials is important, particularly as these businesses seek to grow their market share. A classical case is the Eastern Cape situation mentioned earlier on. Developing a solution that will create dependence of the community to other provinces or even regions in the world is not a good solutioneering approach. But take deliberate efforts to establish the abundance of raw materials that can be used by the communities to produce safe, sustainable, durable modes of transportation begins to solve local problems while creating industries to services other provinces, regions and global south communities.

JT: Much discussion has taken place locally and globally on increasing the global south's share of the global value chain. How can technology empower African countries and other less developed countries to gain a bigger share of global value chain? Take this from the design work at Ditsela Place to actually full life cycle, so Africa gets more value from the global value chain instead of continuing where their primary role is resource extraction for the west and their industries.

KM: Ditsela Place is a new research, innovation and commercialization hub that houses the Gibela Research Chair, the DSI-NRD SARChI in Future Transport Manufacturing Technologies, NRF Flagship Community Engagement program, the Rail Manufacturing Centre for Entrepreneurship Rapid Incubator to ensure an ecosystem that brings value to government, communities and industry. Recently the Institute of the Future of Work has been launched and the technology hub as one of the verticals will be based in Hatfield. In this context this is a relevant and very interesting question because an EU-SA dialogue addressing the issue of localization is a matter that is receiving attention through my Research Chair. We realize localization is more than an issue of creating jobs for people today. It is more about long range planning for getting more out of materials from extraction to processing to production of products and services. A good example is how to localize the production of commuter trains. The invitation of industries that process production of steel to discussions with extractors of iron ore to those that require high-quality steel. There are jobs created along the way in all of these industries. However, we can deal with this situation in silos. The relations between iron ore extraction, steel production and use of steel in products such as commuter trains is critical and requires extensive collaboration. How do we deal with communities that relied on mining and given the mining industry has declined? In most cases the mining town dies. How can we prevent this? Ditsela is examining how to develop new industry and jobs for these communities. How do we address this issue with existing mining towns? We must assist them in addressing the development of industries that complement the mining efforts and allow for expanding the communities even when mines shut down. Substantial assessments of what is the next economy going to center around is a key aspect in the reindustrialization of these towns.

JT: What role does manufacturing play in processing of Africa's raw materials? It would seem processing our raw materials through different stages of manufacturing would increase revenues to Africa and create more jobs. What kinds of technology and technology policies would speed up this process?

We are concerned with the technologies and techniques that will empower communities, create sustainable employment and assure environmental security. What is the responsibility of government in directing manufacturing and transport development to address this?

KM: The government is very important, most of the time even with procurement of materials and technology from overseas the government has a fundamental role. Often the communities are not taken into consideration. Governments must put policies in place that direct the import of technology when needed, but deal with promotion of technology locally. The engagement of communities as stakeholders needs to be improved by states especially in the global south. The finalization of decisions often results in solutions that fail to resolve the social challenges of communities as they often come as an afterthought but not integral part of the development of a solution that must speak to their challenges. Often external sourced service providers are not providing full disclosure since their goal is to maximize their profit. Governments must be aware of this and be very deliberate in how they counteract this external interest. They should work closely with organization such as INAT that can provide valuable guidance in the selection of technology and development of technology policies.

JT: One position is that international agencies have a serious responsibility in addressing global inequality and that industrialization of countries in Africa will significantly reduce inequalities between countries. This requires swiftly advancing the manufacturing sector in these countries.

The work of NGOs and international institutions are reshaping much of what is developing in many countries in terms of industry and services. How can we assure that these non-profit institutions serve the interest of the local communities and the poorer countries?

KM: International agencies have a neutral posture when it comes to addressing issues. Especially if there are independent of influence from the main supporters. An important example in this regard that is not naïve to the fact that there is always interest in a majority of the agencies but those that are nonaligned needed to be engaged as they do not drive external interests.

NGOs are able to provide a neutral platform wherein various stakeholders from developing nations can both be heard, educated based on their actual needs and solutions co-created with experts willing to align their expertise to the social context where development is needed. The efforts around arranging conference and education people is important. One example is the tech-a-thons that INAT has conducted at the various ICATs. This type of work can be extended to working directly with communities. This can go a long way with community engagement to generate community interest.

JT: starting at early stage implies this should be part of the educational process, starting at early age and to the universities.

How can universities reshape curriculum and research agenda to improve the manufacturing and transport situation in South Africa?

KM: From an early age, pre-school even, the articles of play they are given to play with should speak to developing a creative mindset. Everything is evolving. The creation of the flexible creative mindset is so important. This environment generates more creativity. Schools are associated at early age on the technics of writing. The learning process is more extensive. The increased focus on making and breaking things generates more creativity. At higher level lines between work and school must be blurred to engage students more with challenges that speak to where they come from but are engaged with real world problems. They will be able to be continually engaged. The capability to 'make' is very key to where the continent is, and must be made common ground for learning for everyone. Whether we are talking about making commuter rail trains, building a house, or making the bag that will be used to carry the learners' schoolbooks, all of this is essential to help us move forward with appropriate technology in the context of our global village.

JT: One position is that international agencies have a serious responsibility in addressing global inequality and that industrialization of countries in Africa will significantly reduce inequalities between countries. This requires swiftly advancing the manufacturing sector in these countries.

Manufacturing and transport industries are dominated by private ownership. There should be an appeal to private owners, small, medium and large to assure an accountability that reduces inequality and promotes an improved quality of life for a broad population. Balance responsibility of private sector like Gibela with international agencies such as UN and AU such that private manufacturing sector is held accountable to assist in community development.

KM: This is a very challenging issue. International agencies are accountable to agencies and governments that provide their funding. So while they may be responsible in advancing the interest of local communities, they realise they may bite the hand that feeds them. This is a real conundrum. This is why we are having the series on dialogues around localization with EU, private companies and not-for-profit agencies. On one hand with companies like Gibela, but also we have international organizations concerned with what happens in this space and then we have parastatal organizations. If international organizations are not willing to play a strong role the improvement to the communities cannot be achieved. UNESCO, two European countries and five African countries collaborated to address community based design to come up with solutions that take into account resources in community to take mobility project to production. With different partners taking into account the importance of collaboration. Most consider how the local communities solve their problems while not relying on dependency on these external collaborators. The creation of dependency is limiting on how the community can advance in terms of self-empowerment.

JT: good way of taking us to the final point.

In the final analysis, the individual citizen and their community organisations will have to take on the responsibility of changing the manufacturing and transportation sectors to improve the well-being of citizens and guarantee the ecological sustainability of their communities. In final analysis what do you think the citizens can do to improve the situation and foster a climate with more AT in general

KM: Considering the work that the Research Chair is doing sustainability is a key driver in all its work. Even if one gets support to start, sooner or later one must be self-sustaining these activities to continue into the long term. As Industrial Engineers the goal of the profession is to implement efficient systems to transform inputs into outputs in a process. In manufacturing scientists and practitioners look at this cyclically . What is the life cycle of the manufacturing project. However, collaboration is very important. The collaboration among African countries is very critical. Collaboration of South Africa with countries outside of Africa is also important. We really collaborate with others in East Africa or West Africa. It starts with communication. Too often there is a chasm in talking with each other. The different conferences have proven as one way of increasing the communication and sharing information on appropriate technologies. Sustainability comes primarily comes from investing in the individual that understands the nature of AT, AT by its very nature is about sustainability. It means the resources are local and available for the construction of the AT. The current modes of transportation are an example of how we are often not using AT and therefore creating problems when it comes to achieving sustainability.

JT: Those that are involved with INAT have more knowledge about AT and therefore are more responsible for its development and spreading. What about the non-technical citizen? For example how does your community-based farming project link in to what is needed? They are locals that are not engineers, they are an example of average citizen. How do you get them mor attuned to using AT.

KM: One principle for us is, even though we are in advance technology space, we always must remember where the people are. If you understand that context, you will always keep in mind the importance of AT. In this environment you will keep in mind the issue of scarcity. With regard to the food project, the solution on develops will address the importance of basic daily needs. It this case food security. In all environments we have to examine the issue of scarcity. It is easy to lose sight of the realities of day-to-day African people. Researchers must find ways to stay connected. In this project there is a local school nearby and finding a keen interest on the part of the youth and their development premised on real day to day issues becomes transformational.

JT: This is good point to conclude our conversation. We end by reflecting on the community and the youth. Any final comments as NRF SARChI chair in Future Transport Manufacturing Technologies

KM: From AT perspective, there is a lot of talk about moving away from diesel and petrol vehicles, but we must look at where our communities are in terms of addressing the transportation needs of our people. We need to listen more to our communities in all matters of technology advancement and then the interventions will be most appropriate.

JT: I just read in Car magazine about a comparison of Petrol and Elective cars. The article indicated that an electric car had to travel 200,000 km to break even in terms of GHG emissions because of more GHG emissions in manufacturing of electric car due to battery and electric motor production. An additional problem occurs if the electricity is produced by carbon intensive means such as the coal-fired plants in SA. That is a whole other story – appropriate technology needed in the energy sector.

CHAPTER 7

Appropriate Technology and Mining

Thomas Dalgety

The Cutter-Head Dredge

Technology at the workers level is machinery, large industrial plants, and equipment. The goldfield workers in Guyana's hinterland call the cutter-head "Brazil technology" because this equipment was 'Made in Brazil'. It is a Brazil export hence it is Brazilian technology. At the learned level of sensibility however technology is the application of scientific learning. This learning of science & technology includes design and mathematics. Thus, for the learned in Guyana, the cutter-head attached to a river dredge is a mechanical cutter with an electric motor attached to it, designed to disintegrate compacted gravel on the bed of a river as a suction pump takes the gravel from the bed of the river to the surface of a river. The cutter-head dredge is for mining ores on gravel sections of a river or port or otherwise.

Mining is concerned with the responsible extraction of minerals from the crust of the earth. Early mining efforts were distinctly linked to conquest and the near extermination of indigenous peoples of the Americas.[1] The minerals are mainly compounds of the elements, metal ores and silicates. Many of the metal ores are oxides (Aluminium Oxide), sulphides (Zinc Sulphide), chlorides (Sodium Chloride) and carbonates (Calcium Carbonate). Mining therefore uses the science of chemistry, physics, geology and engineering to extract these metal ores from the ground and extract pure metals from tons of ore. Responsible mining is sustainable and admits that the ways of extraction of finite resources must consider the effects on renewable biological resources. Careless extraction processes can lead to extinction of species and all species should be allowed to coexist in our massive ecosystem. However, good mining policies and strategies can lead to optimal utilization of finite and renewable resources and national wealth. Good policies impact the standard of living of all ethnicities in a nation and attempt to remove the distinction of rich and poor. The rich/poor barrier will only be removed if the economic table seats all ethnicities with the will to maximise sustainable use of every natural resource.

In Guyana much focus is given to gold, diamond, bauxite for export to foreign lands with little attention given to the aluminosilicate range of minerals for rapid local industrialization of the country and sustained comfortable livelihood of a majority. The aluminosilicates, in the wider world, are used to manufacture a very wide range of commodities that contribute to comfortable living. The wise management of flora and fauna must never be overlooked. Recently global warming conditions caused a range of strange climatic conditions, hostile to many species. Man must be ready to intervene and save flora and fauna. Residents of one of our hinterland villages were able to save hundreds of turtles by digging up the eggs, burying them and releasing the newly hatched turtles into their habitat. As ethnicities of the world moves forward, the united focus must be to help each other apply design, development, articulation and implementation of policies, technologies and education so that together we grow. Ethnic collaboration must replace competition and rivalry. Integrity must replace greed and corruption. Let all learn to develop all our resources in a responsible manner.

Stakeholders and Manifesto

The International Network on Appropriate Technology (INAT) seeks to publish an Appropriate Technology Manifesto to impact technology policy and the acceptance of appropriate technology by 6 stakeholders. These are 1) educational institutions 2) government structures 3) international agencies 4) NGOs and non-profit organisations 5) private sector Companies, and 6) engaged citizens. This will be outlined for mining in the order given.

Education Institutions in African communities around the globe must immediately support instructions and learning of elements of the universe and their ores. This will bring about development and wealth in these communities. Salt is legendary in Africa. Many legends are told of salinity of rivers and manufacture of salt in West Africa. Salt trade across the Sahara is legendary. In the African diaspora the Caribbean Island of Anguilla is nothing but salt. Yet, African education institutions don't point this out to students nor instruct students to make to sodium hydroxide and sodium bicarbonate for African communities. Bauxite in African communities around the globe is mined and shipped to other ethnic communities and not transformed in situ to made refractories. So is clay, kaolin, iron ore and compounds of the elements silicon, phosphorus and sulphur mined and shipped to other ethnics.

"Why are we so rich and yet so poor?" asked the Black Think Tank in 1992. Education Institutions in African communities do not deliver inquisitiveness. Be inquisitive about limestone for African communities, bauxite for African communities, sand and silicates for African communities, phosphates for African communities, Sulphur and sulphuric acid for African communities, chlorine for African communities. Mining education will lift minds. The Periodic Table must be learned simultaneously with the Alphabetic Table.

Government Structures must stop regarding Africa and African communities as a football to be kicked around. Government structures must treat mining as paramount. Of the Russia Revolution, Lenin wrote the "One miner is equal to 200 peasants in the ability to act historically". African communities must demand scientific support in times of disaster. Where there is feudalism government structures must show daring. Land titles must be shaped to the advantage of natives so as to give them the opportunity to grow from artisans to large establishments siting at the economic table of the world.

International Agencies such as Extractive Industries Transparency International EITI and Conservation International IC must provide funds for appropriate technology in developing countries. Funding enterprises doing appropriate technology must be their mandate.

African NGOs and non-profit organisations must mushroom to match Indian, Chinese and European NGOs. Ethnic NGOs are fierce supporters of their communities. In Guyana, the Maha Sabha and Indian Arrival Committee are two Indian NGOs that put mining and minerals on their agenda. They lobby all governments on mining on behalf of the Indian community and many Indian importers of engines and equipment for gold mining now have mining property. They also have access to funds when Lottery and Auction of property with gold is executed by the government. The Appropriate Technology manifesto advocates creating NGOs that will lobby for appropriate technologies

Local Private Sector Companies must be the watchdogs on the behavior of international companies. For private sector companies to take the mantle of extraction of ores from International Companies they have to build technology that is appropriate to their environment. Private sector companies must market and sell minerals mined by their fellow citizens. Profits motivate private sector companies and the bedrock of greater profit is appropriate technologies for larger production.

Engaged citizens are the lens to the future. Engaged citizens must innovate and invent. This takes place when interest changes to attitude. It is the attitude of citizens that produces wealth. Engaged citizens are also patriots.

Appropriate Technology for Mining Past, Present and Future

Appropriate technologies can promote efficiency and effectiveness in any endeavor. However, without ethnic leaders of integrity in all sectors, wealth gained through use of appropriate technologies would not automatically result in reduction and eventual elimination of squalor and degradation of the living condition of the African working classes. European nationals used natural science and appropriate technologies to deplete other ethnics of ores and enrich themselves and their nations. Education for the oppressed peoples focused on content and skills that allowed them to stay in their stations. Some who were singled out for small compliments and honours considered imported technologies as innately better than the local technologies although many of the oppressed fabricated any part of any machinery necessary for the efficient functioning of the European factories. Factories in colonies worked nonstop. Factories in colonies did not stop for broken parts to be imported. Local blacksmith enterprises could always be depended on to repair or fabricate what was required.

A few smiths and technicians were invited to 'higher education' in Europe but many of them remained in Europe to contribute to Europe's development. Some of those who returned to their native land did not adjust to local conditions and needed to use the same equipment that was available in Europe. Professionally trained graduate science teachers in Guyana demanded models and overhead projector slides from overseas to teach the development of the frog and butterfly even at the primary level which suggested that their exposure had not prepared them to connect with their homeland students. These teachers could not induce change by relating with the reality of homeland lives. Many of those who returned to head homeland industries also had problems adjusting. They too wanted the European technologies they used abroad. They could not appreciate that local staff could improvise and solve problems. They could not accept them as persons and trust them enough to have them contribute maximally to the growth and development of the enterprises. The result was that whereas during colonial times factories worked continuously stopping only for planned maintenance, in the Independence era factories spent much time idle, waiting for parts to be brought in from overseas. In many cases new machines were imported because the 'education' of the top management did not prepare the team members to consider what was appropriate to the needs.

In Guyana, the Guyana Geology and Mines Commission, GGMC, is the government authority charged with the responsibility of managing the minerals and mining sector of the national economy. Although bauxite is and manganese ore was also mined in Guyana this paragraph deals only with gold mining. It attempts to show the difference that policy change and regulations can make in empowering individuals who then contribute to national wealth.

The Guyanese in the bauxite and manganese ore industries are mere employees. During colonial times very harsh laws greatly inhibited the ex-slave pork-knockers from making a sustainable living[2]. Officially they were granted twelve-month privileges to mine with battel and jig. The pork-knockers with privileges were called tributors. Tributors were allowed to keep part of their earnings as raw gold. The tributor found by the police with raw gold but no privilege could be jailed. The mining landscape then had only large companies and tributors.

During 1991, the Mining Act of 1989 which nullified that of 1921 became operational. The new regulations allowed the Small-Scale miner to operate a property claim of twenty-seven acres land or for a river claim up to one mile of navigable river. These regulations catered for a new group, Medium Scale miners, who could work on properties of one hundred fifty to one thousand two hundred acres. The 1991 regulations also catered for Large Scale miners to work on properties from five hundred to twelve thousand eight hundred acres. Guyanese wanting to start in mining can start with only the Guyana equivalent of five United States dollars. This allows someone to buy a sixty-day prospecting permit. Once he finds gold, he can apply to the GGMC for a claim. Since 1991 more people applied for claims annually. The GGMC sets production targets which are often surpassed by the miners. The Guyana gold mining landscape includes very many Brazilians working in partnership with Guyanese medium scale miners. The Brazilian miners are responsible for the introduction of equipment like the cutter-head used in dredging. Hard rock mining is also done in Guyana mainly by a family in the gold mining business for nearly a century. Their involvement is sustained by their use of appropriate technology.

A few years ago I had the privilege of interacting with Sudanese officials aiming to increase revenue from gold mining since foreign royalties from oil extraction had declined. The areas on the rivers identified for mining were well mapped. They were North Sudan near Lake Luba, Abuhamad on the Nile, South Khashn Algirba on the Abara River and Ed Damizin City before Roseris Dam on the Blue Nile. The Sudanese artisanal miners are equivalent to Guyanese pork-knockers. The pork-knockers are the first inquisitive prospectors of gold throughout the hinterland. They grapple with the problems of designing systems including machines to extract the gold. Since 1991, GGMC staff sample what the pork-knockers dig up. Unfortunately, Sudan artisanal miners are unlicensed, unregulated, and not allowed to use machines. They dig holes in the ground; treat the silt with mercury then heat to extract gold. The Sudan artisanal miners need to be regulated and empowered. Their desire to succeed will kindle the inquisitiveness to find systems, including machines, which work effectively and efficiently. In Sudan the small-scale miners have a mining lease of one square kilometer and are licensed to mine alluvial gold, chromium, mica, gypsum and the aluminosilicates. Large scale mining is done by non-nationals. They begin with concession agreement with the central government, produce a feasibility study then do a joint venture. In my travels in Africa, I have not seen Africans mining with machines that match the land titles. Success is maximized when appropriate technology is used to match land titles

The goals of the African Union Agenda 2063 are bold and the strategies, including the Science, Technology, and Innovation Strategy for Africa 2024, taken to achieve them are commendable. Sustained and responsible mining could contribute greatly to realizing these goals.

Mining has a legacy of violence associated with it. From the coal mines of West Virginia in the USA to platinum mines in Marikana South Africa workers have faced armed suppression. 14 August 2022 marked ten years since the confrontation at Lonmin mines in Marikana left 34 dead and 78 injured [1]. Violence when workers confront management is not unique to mining. However, history shows mine workers have faced a higher level of violence when they demand more. This inconsistency is linked to the rationale for and conditions of mining operations, which in turn are linked to the technologies of mining operations. Mining represents the most basic conquest of Mother Earth. The centuries old practice started with hand tools digging rocks from the surface or from caves. Today, powerful explosives blow mountaintops away, and a complex of machinery digs miles into the ground extracting millions of tons of earth[3]. Mining's destruction to the landscape is obvious with gaping holes, piles of tailings and toxic fumes over miles and miles of landscape. The damage is much deeper to both planet and people. Seismic activity, sinkholes and groundwater contamination are all common in mining areas. Across the planet, mine workers have suffered death in mining accidents as well as respiratory illnesses (many resulting in death) due to toxic poisoning. "Although the mining industry employs only 1% of the global labour force, it generates 8% of all fatal accidents" [2]. South Africa finds itself in a particularly precarious position. Under apartheid, mining was developed as the major revenue generator with a system of grave inequality and endangerment to workers. Today, the government, private sector, citizens, and all other stakeholders are faced with the task of redoing this sector with humanism at the core. This requires engaging appropriate technologies coupled with policy changes and community support [3], [4], [5], [6].

Conclusion

Africa is unique, special. It is the only continent that stretches north of The Tropic of Cancer and south of the Tropic of Capricorn, allow it to experience a wide range of climatic conditions. It is believed to be the cradle of civilization where human life originated. It is the home of the Nile and Congo among the largest rivers of the world; about two thirds of earth's arable land; the Sahara; and a wide range of minerals. The Democratic Republic of Congo (DRC) is regarded as the most mineral rich country in the world. Many other countries have considerable deposits of minerals. Africa is unique, special region. Colonialism with the Berlin Conference in 1884-5 divided Africa into more micro-states than any other region in the world While collectively Africa is resource rich, the divide-and-conquer tactic continues to allow exploitation of Africa's minerals.[4]

What is the purpose of this Appropriate Technology manifesto? Religious and cultural councils must become educational institutions and make African communities conscious and inquisitive about mining and minerals. Government Structures must prioritize mining and minerals for all ethnic communities. Africa must not be left behind for criminality to make the world unsafe. International Agencies must fund all ethnic communities and bring equity to the table of economics and wealth for all humans and so destroy poverty. Ethnic NGOs and non-government organizations must cooperate and collaborate in the mining and minerals sector to prevent wars at the racial level. Private Sector companies must function as sellers and marketers of ores and minerals produced by artisan and small-scale miners and so contribute to appropriate technologies. Engaged citizens must be lens of the future.

Bibliography

[1] I. Mahlangu, "Little has changed in public order policing since massacre," *Sunday Times,* p. 15, 14 August 2022.

[2] Howden, "Howden," 5 December 2019. [Online]. Available: https://www.howden.com/en-us/articles/mining/we-need-to-talk-about-mining. [Accessed 14 August 2022].

[3] D. Gqubule, "Transformation in South Africa's mining industry," in *The Future of Mining in South Africa Sunset or Sunrise*, Johannesburg, MISTRA, 2018, pp. 121-143.

[4] R. Harvey, "Greening South African mining through the Fourth Industrial Revolution," in *The Future of Mining in South Africa Sunset or Sunrise*, Johannesburg, MISTRA, 2018, pp. 144-178.

[5] L. Fioramonti, "The future of mineral mining," in *The Future of Mining in South Africa Sunset or Sunrise*, Johannesburg, MISTRA, 2018, pp. 358-380.

[6] H. Deedat, "The mining-energy nexus, climate change and prospects of just transition: Reflections of a labour educator,"

in *The Future of Mining in South Africa Sunset or Sunrise*, Johannesburg, MISTRA, 2018, pp. 335-357.
[7] O. Pert-em-Hru, Pan-Afrikanism The Battlefront: Afrikan Freedom Means Defeating Neo-colonialism, London: Omowale Ru Pert-em-Hru, 2017.
[8] D. Graeber, Debt: the first 5,000 years, New York: Melville House Publishing, 2011.
[9] M. Hart, GOLD The Race for the World's Most Seductive Metal, London: Simon & Schuster, 2013.
[10] NS Energy, "NS Energy," 30 June 2021. [Online]. Available: https://www.nsenergybusiness.com/projects/mponeng-gold-mine/. [Accessed 14 August 2022].

Endnotes

1 "Then came the conquest of Cortes and Pizarro. Between 1520 and 1640, untold tons of gold and silver form Mexico and Peru were transported across the Atlantic and Pacific in Spanish treasure ships. ... Christopher Columbus-a Genoese mapmaker seeking a short-cut to China-touched land in the New World, and the Spanish and Portuguese empires stumbled into the greatest economic windfall in human history: entire continents full of unfathomable wealth, whose inhabitants, armed only with Stone Age weapons, began conveniently dying almost as soon as they arrived. The conquest of Mexico and Peru led to the discovery of enormous new sources of precious metal, and these were exploited ruthlessly and systematically, even to the point of largely exterminating the surrounding populations to extract as much precious metal as quickly as possible." [3, pp. 309-311]

2 The term Pork-knockers is a nickname given the miners after the pickled peccary, or wild hog, they often eat.

3 The conquest of the planet in search of gold has been particularly devastating. The plunder of South Africa is particularly significant. The Mponeng gold mine is located forty miles from Johannesburg. It is the deepest man-made hole on earth and currently the deepest operating mine with current digging four kilometers below the surface. It takes the processing of four tons

of orc to get one ounce of gold [5]. "The town of Springs lies on the Witwatersrand, the richest goldfield in history, a 300-mile crescent that arcs around Johannesburg in a wide belt of gold deposits. Forty percent of all the gold ever mined in the world has come from that single geological formation." [4, p. 12]

4 "Afrika is the richest continent on earth. It is not just marginally richer than the other continents, it is massively richer. The Afrikan continent is blessed with an abundance of the world's most precious minerals and in many cases has more reserves than the rest of the world put together. ... It is the sheer magnitude of the riches contained in the soil of Afrika particularly the Congo, coupled with the greed of power grabbing European imperialists that eventually led to the 1st and 2nd world wars." [2, pp. 29-31]

Chapter 8

Appropriate Technology and Agri-business

Brian A. Stephenson

In this chapter we look at appropriate technology through three pathways to sustainable agricultural development: Ecological Restoration, Food Sovereignty, and Regenerative Agriculture. With more than one-half of the world's population now living in cities, meeting the food and water needs of these ever-expanding urban centers has over-stressed the source of food and water supply: the agricultural sector. Failure in the agricultural sector is a trigger to unhealthy demographic shifts from rural to urban centers, further increasing the demand and depleting the supply for food and water. The present reality of global warming, water deficits, conflicts, chronic poverty, and food insecurity threaten the very survival of life. What is the sustainable way out?

In charting a sustainable course, we can invoke the Akan notion of SANKOFA: to realize a sustainable future, we must look back, learn the lessons of the past, and act accordingly. The present indicators are that the global ecosystem has failed. The failure has disrupted the availability of food and water. We then ask, "What are the causative factors for our global and local ecological system failure, and what must be done to restore it?" By rebuilding the ecosystem, we will strengthen our broken economic and social systems. Regenerative agriculture is introduced as an agricultural system that recognizes that the soil is close to dead and needs revival. Healing the ecosystem is priority one.

We next examine food sovereignty. Who controls our food chain? Food sovereignty demands that the hundreds of millions of small farmers, fisherfolk, pastoralists and indigenous people who produce food, have a say in the quality of their food chain, not the large agribusinesses or food retailers. Appropriate technology is a people-first approach to innovation and problem-solving that sustains the next generation and protects the planet. It reaches back to the knowledge of the ancients but also telescopes into the future, integrating disciplines such as information systems, biotechnology, microbiology, and ecology. Once informed, sustainable solutions and innovations can be unleashed. Placing the planet and people first leads to high, long-term agricultural productivity.

Sankofa

In order to realize a sustainable future, we must look back, learn the lessons of the past, and act accordingly. This belief has a name in the West African region of Ghana. It is "Sankofa!" Its image is of a large bird facing forward but looking backwards. In our often siloed society, engineers and scientists can be uninformed of the negative role technology may play in our social and economic frameworks. They rarely wrestle with the ethical implications of their work or examine the social consequences of its impact. The customary environmental impact report is sent to senior managers and policymakers whose decisions may be based on politics or financial incentives. So too, the leaders of industry and government! Their election to leadership weighs character, experience, and personal ethics; not the global implications of their decisions on society or environment. Sustained leadership in industry is grounded in sustained corporate profit. In the government sector, elected leaders court voters based on sensationalized popularity. Their policy positions often are tied to who gave how much to their campaign. It is a rare leader who learns from the past to define a better future.

To what extent is the food chain in the developed United States under the sovereignty of American farmers? A comforting statistic informs us that 97.6 percent of all U.S. farms are family-owned farms (less than 5 acres) [1]. We can then ask, "what is the average plot of farmland in America?" A 2013 U.S. Department of Agriculture study documented the midpoint acreage of U.S. cropland between 1982 and 2007. In 1982, the midpoint acreage was 589 acres (one-half of all farms have higher acreage than the midpoint), an already very high number. By 2007, the midpoint acreage of U.S. cropland had nearly doubled to 1105 acres [2]. Fourteen years have passed since that data point at the time of this writing. The trend has continued. Corporate control represented 2.4% of U.S. farms. When we do the math based on total acres of farmland owned by the 2.4%, we conclude that 99% of all U.S. farmland is corporate controlled. American farmers have very little influence on the food chain. Unbridled corporate power and political influence allow food corporations to control the actors in America's food chain and to manipulate market prices. The power of corporate agriculture is global and their contribution to the poverty equation is high.

Of the big U.S. farmland owners, standing tallest is Bill Gates with over 242,000 acres in 19 states. Their non-profit, Ag One, promises to empower smallholder farmers in sub-Saharan Africa and south Asia with the affordable, high-quality tools, technologies, and resources they need to lift themselves out of poverty. Unfortunately, their pathway to accomplish this is through "inappropriate technologies "which are input intensive and dependency motivated. Most recently, at an annual African farming summit in Rwanda, activists, farmers and faith leaders demanded the Gates Foundation and other funders to stop supporting an effort they say has failed to deliver on promises to radically reduce hunger and increase farmer productivity and income. The cost of expensive seeds, fertilizers, and pesticides place farmers in a form of bondage.

An abundance of cheap labor and unrestrained access to raw material has been the foundation of industrial progress for over millennia. The continent of Africa has always been the source of both. The wealth and global dominance of the so-called developed nations has been inextricably tied to Africa. For them to take the proverbial "hand out of the cookie jar" is not just an ethical dilemma; it can cripple their national economy. Any serious attempt to cut the global north's umbilical cord from mother Africa is a threat.

Nature has limits. Relentlessly extracting oil from the earth, a resource which reportedly took over 300 million years to form, has global consequences for present and future generations. In the agricultural sector, massive logging and mining, large scale monoculture farming, and commercial fishing are behind the litany of climate change impacts. Colonialism and global capitalism have generated an enormous ecological debt to the world's citizens.

The decision-makers, the leaders of industry and government, are long-term influencers for short-term gain. The Sankofa principle is largely absent when development priorities for people groups and nation-states are being established. Learning from the past is relevant only in the short-term. It is preamble, relegated to the margins of academia and personal conviction. Knowing what is better for the planet does not incentivize a change of direction.

Ecological Deficit

Ecological debt is the debt accumulated by aggressive countries for plundering other nation-states, exploiting their natural resources, degrading the natural habitat, beggaring or enslaving the population, and/or freely occupying their ecosystem to discharge waste. The charge of ecological debt against the world's wealthier countries is an implied demand, by the impoverished countries, for fair recompense. Ecological deficit recognizes that, regardless of who caused what, there are present imbalances and dysfunctions in the ecological systems that culminated in high poverty for that nation-state. Land exploited is less productive. Flora and fauna that once proliferated with purpose, are no longer. Human coexistence within these ecosystems can no longer be supported. Ecological damage and injustice requires ecological healing. Healing the ecological deficit is the pathway to economic rebirth and resilience.

The Not Green Revolution

Most of the world's poor rely on agriculture for income and sustenance. Globally, there is enough food for everyone. Yet, hundreds of millions of people are unable to grow or buy food sufficient for their physical needs. Many cannot grow crops in environmentally safe ways! This reality is everywhere but is more evident in Africa. Agricultural production for the continent is increasing at slightly more than 2 percent per year, while the annual population increase is greater than 3 percent annually. Farming is becoming simultaneously less productive and more damaging to natural resource biodiversity as most production increases have come from taking more land into cultivation. The whole continent of Africa currently produces less maize than the state of Iowa; primarily due to Africa's lower productivity rates. Access to water is a significant factor. African soils have declined in fertility over the past 30 years because the ground has been repeatedly farmed without the adequate replacement of plant nutrients. Farmers in Eastern and Southern Africa apply 10 to 15 times less nitrogen fertilizer to crops than their counterparts in developed countries. Fertilizer costs can be as much as three to five times higher than those seen in the United States or Europe, which is a barrier to cash-strapped farmers. African average maize yields continue at around the same low levels they have been for at least 30 years. At the 2006 African Fertilizer Summit in Abuja, African Heads of State reiterated that solving fertility challenges is the catalyst needed for an African Green Revolution [1]

The Green Revolution strategically consolidated millions of small and medium-scale farms into the hands of a handful of global agribusinesses characterized by plantation-type, machine dependent, monoculture "modern" farming. The Alliance for a Green Revolution (AGRA) persuaded and promoted the indiscriminate use of pesticides and chemical fertilizers disrupting the ecological health balance in humans, wildlife populations, and plantlife. The excessive reliance on synthetic fertilizers, and the improper use and disposal of animal wastes consistently make members of this alliance the top violators of the U.S. EPA Clean Water Act. The inappropriate farm management practices that are promoted increase soil erosion rates, resulting in the loss of productive farmland and contamination of waterways. The water irrigation demands of mega-size farms continue to suck many natural freshwater resources dry, increasing the salinity and toxic mineral levels of these resources. AGRA is global and where environmental regulation is weak, the violations are more grievous.

Timothy Wise, senior adviser at the Institute for Agriculture and Trade Policy noted "that there is no sign of a productivity revolution in AGRA's focus countries — nothing that is significantly raising incomes and reducing hunger." [2] His research group wrote that "instead of halving hunger, the situation in 13 focus countries has worsened since AGRA was launched. The number of people going hungry has increased by 30 percent during the AGRA years." Farmers became impoverished and exposed to environmental risks. "The introduction of genetic engineering - the driving force behind AGRA initiative - will make smallholder systems more environmentally vulnerable in Sub-Saharan Africa. It will also lead to the indebtedness of these farmers."[3] The food sovereignty movement was born to counteract AGRA's domination.

Mining in South Africa

The parent of South Africa's economic wealth has been mining; at one point mining represented 20% of their GDP. The offspring of mining wealth in South Africa has been its robust industries: agriculture, finance, transportation, and military. Prior to the end of apartheid in 1990, whites and coloreds who represented 20% of the population, controlled the entire economy. Indigenous blacks working in the mines for little or nothing were the life blood of the lucrative mining industry. The social cost to this labor force included the disintegration of their families, subjugation of their cultural social systems, and their inability to control the food chain.

The end of apartheid was great politically but how has that change benefitted the oppressed South African blacks? In reality, the economy has not changed much for many. "A multifaceted inquiry into the state of South African economy that includes all these forms of capital leaves no doubt that white capital continues to dominate the economy"[4]. South Africa today exhibits the most unequal income distribution in the world. Most significantly, and least surprisingly, this inequality remains based on race. Eighty-five percent of Africans remain poor, while 87% of whites are in the middle- and upper-income brackets. Since 1993, the percentage of whites in the highest income levels increased from 8% to 20%. At the same time, the percentage of blacks in the middle class increased from 7% to 14%, with only 1% of blacks reaching the very highest income levels[5]. The integrity of the food chain that was dysfunctional for the black population prior to 1990 has not improved. The negative environmental impacts of the mines on the ecological balance have grown, not diminished.

Political control is not economic control and economic control dictates agriculture. South Africa's economy during apartheid was structured so that industry-generated revenue was split between South African whites and investors overseas with little left for the black population. The 700 state-owned enterprises under the Department of Public Enterprises have a direct link to the center of capital accumulation, the mining industry. As the chairman of Anglo American said at the time: "Far, therefore, from secondary industry offering an alternative to take the place of mining in our economy, the fact is that the prosperity of gold mining and the other primary exporting industries is a prerequisite and a condition of our industrial development. ... Manufacturing industry will only be able to serve as a substitute for the wasting asset of our mining industry if it is able to increase substantially its contribution to our export trade."[6]

Unresolved problems in South African mining, particularly on gold mines, are enmeshed within the system of production through mining methods and labour practices entrenched by apartheid. This system sets the parameters for, and hence limits, strategies designed to improve the ecological system that were damaged. Mining destroys biodiversity, which reduces agricultural potential. Mining destroys, fragments, and degrades natural habitats, releasing toxic wastes and altering land-use dynamics in mined regions [7]. Knowing the past and its continued adverse influence on ecological health does not incentivize a change of direction.

Haiti: A Legacy of Deforestation

Forests provide a wide variety of highly valuable ecological, economic, and social services, including the conservation of biological diversity; carbon storage; soil and water conservation; provision of employment and enhanced livelihoods; enhancement of agricultural production systems; and improvement of urban and peri-urban living conditions.[8] Forests help lower the ground temperature and generate rainfall. The evidence shows that global warming has caused an increase in heavy precipitation events, the kind most responsible for major floods. Haiti is at the top of the list for deforestation. The country retains less than 1% of its original forest cover. There, deforestation and soil erosion have greatly decreased the agricultural productivity of the nation, especially in the once fertile Artibonite Valley. A six-month long drought in 2019 destroyed nearly 50% of the winter season harvests [9]. What were the contributing factors? The devastation of Haiti's forests is caused by a range of socio-political-economic forces. These forces have directly contributed to the nation's dismal economic condition. Slavery had the first ecological impact. Large scale deforestation for sugar and coffee plantations was a dominant disruptor to the island's biodiversity.

After Haiti became an independent republic, efforts to continue working the land for sugar continued but with great economic hardship. The U.S. and France collaborated on economic isolation of the young nation. An 1825 debt imposed by France, referred to as an "independence debt", along with a requirement that France receive a 50% discount on their exported goods, crippled the nation. "Haiti's military victory ended all hope of resuscitating its sugar commerce, already devastated by war. Slaveholding nations, enraged and humiliated at the ex-slaves' audacity in freeing themselves, punished Haiti with commercial embargos and diplomatic ostracism, and France levied crushing reparations. Within the new republic, conflict between black and lighter-skinned Haitians accelerated the economic decline. The collapse of Haiti`s sugar industry and the subsequent shortfall in world sugar markets created opportunities for other sugar producers, notably Louisiana and Cuba, where so many Haitian planters and sugar experts took refuge from the war." [10] With sugar in high demand, the country internationally isolated, and new supply markets operating in Cuba and Louisiana, existing plantations in Haiti were overworked trying to keep up with the debt obligations. Patterns of deforestation only worsened.

Shortly after the 100-year independence of Haiti, the Haitian-American Sugar Company was founded. What followed was another tumultuous period in Haitian politics punctuated with a series of presidential assassinations and coups d'état. After successful manipulation of its 1915 elections, the U.S. government invaded with 3000 marines and took over control of the nation. The U.S. attempted to force a new constitution onto Haiti in 1917, one that allowed foreign land ownership. Other policies were introduced including racial segregation, press censorship, and forced labor.

During the U.S. occupation, all possible arable land was cleared to harvest tobacco, coffee, and sugar cane. Persuaded by the U.S. Government, President Elie Lescot, 1941-1945, led the destruction of large tracts of trees for the experimental growth of a rubber substitute weed known as criptostegia. Haitians claimed the policy to have thrown peasants out of their homes without proper compensation, destroyed the best land and fruit trees in the Republic. It produced no rubber [11]. Lescot also encouraged the Church to wage a vigorous extermination campaign against the practice of vaudou, burning ceremonial objects and drums, and cutting down sacred Mapou trees. The U.S. occupation lasted from 1915 to 1934. The agricultural policies imposed devastated what was left of an already wounded ecosystem. In 2022, some 90 years later, the present economic realities of this nation are still connected to this ecological disaster.

The Dominican Republic and Haiti share the same island mass but any fly over reveals the stark differences in forest cover. Why is there such a disconnect between public policy and the self-destructive behaviors of the populace? At present, fuel for cooking and domestic needs continues to destroy Haiti's remaining trees. "Desperate to survive, Haitians are slowly gnawing away at their last one percent of forest, turning trees in state preserves into lumber, firewood and charcoal and burning the grounds to plant vegetable patches. In Port-au-Prince, thousands of sacks of charcoal and kindling and scores of trucks piled high with pine, avocado, mango and apricot wood for building, cooking and dry-cleaning clog the markets."[12]

Hurricane Ivan and Hurricane Jeanne in 2004 impacted Haiti in a disproportionately severe manner. The mud-laden flood waters of Hurricane Jeanne destroyed more than a thousand lives as the floods converged onto the coastal town of Gonaives. Deforestation is the single-most reason for these disasters. Knowing the contributing factors to Haiti's repeated devastation of its forests has not been enough to change the direction towards sustainability. Forces external to Haiti's civil society readily provide disaster relief while its national policies undermine its sovereignty and capacity to solve its problems.

Land Law

Equitable land access and security stimulate economic development through land investment, agricultural production, sustainable management of natural resources, development of market economies and even income generation from taxation. This has been observed in developed countries with balanced and innovative land distribution reform policies. Conversely, in many emerging economies, access to land by the poor is limited. The limited rights of access to land and other natural resources and the insecurity of these rights often lead to extreme poverty and hunger. As a result, the poorest persons in the world are those that do not have access to land ownership and property rights [13].

In developing countries, one common challenge with land access and security is land tenure dualism. Land tenure dualism refers to the juxtaposition of customary laws and modernistic land-management laws and is at the root of the widespread land disputes in Sub-Saharan African nations. Moreover, it partly explains the high pressure of land demand that the urban environment exerts on the rural environment [14].

In Mali, land is collectively managed under customary rights. The same land used by a farmer can be used by the pastoralists to graze their herds. When the land is sold to a commercial entity, many members of the community lose their livelihoods. In Ethiopia, the land laws are not equitably applied. Certain ethnic groups are discriminated against through arbitrary evictions and land grabbing. A foundational assumption for sustainable farming is that the farmer controls their land. When he/she does not have control, applying appropriate agricultural technologies and embracing regenerative agriculture provides only short-term value. Once land has been regenerated and is fruitful, the resourceful farmer can be summarily removed. This also applies to scenarios where large tracts of land are managed by wealthy urban owners. The farmer is merely a tenant with little incentive to invest in sustainable productivity if the owner has the right to absorb the economic benefits of their smart work. In Zambia, more than one-half of the farmland is medium-scale and fits this profile.

Agrarian land reforms should be pursued in broader frameworks of development initiatives, and these frameworks must be comprehensive and integrated to function proactively for the poor [15]. They should identify and target appropriate sub-groups and apply forward-thinking strategies to achieve the intended poverty alleviation outcomes. Land law reform is also necessary but the political will to implement drastic changes can be countered with fierce resistance. Technology and land law indeed intersect. Devel oping appropriate technologies to meet the needs of impoverished agrarian communities must consider land law and encourage civil society to mobilize for change.

People Displacement to Cities

The African continent has the highest rates of urbanization in the world. Because of this, cities are the largest and fastest growing agricultural markets in Africa, with between $200 billion to $250 billion per year in food sales. More than 80% of those sales come from suppliers on the continent [16]. As urban populations swell, the rural (agricultural) populations diminish. The influx to cities is predominantly coming from within the continent. The dietary choices of middle and upper-income consumers have a higher carbon footprint and are more land-intensive. Demand for imported food products in Africa may be a reflection of the appetites of wealthier urban dwellers and the foothold in the marketplace retained by colonial-bred businesses.

Globally, the United Nations projects continued increases in the urban populations to a 2050 level of 68%. Close to 90% of this increase would occur in Asia and Africa [17]. It is also likely that the proportion of the global population not producing food will continue to grow. The pressure for food supply by these ever growing cities are on the backs of the rural sector.

Quite often, shifts in population from farm to city are precipitated by global forces beyond the reach of the community. Haiti again comes to mind. Under the U.S. policy of "trade liberalization", the Haitian government was persuaded to allow rice from the U.S. to enter the country at a low-cost point. That cost point was lower than what local rice farmers could survive on. According to Oxfam, Haiti was virtually self-sufficient in rice in 1980. By 2010, it was importing some 80% of its rice and 60% of its overall food supply. "The foreign rice that is "dumped" in Haiti therefore exacerbates the rural-urban drift that has seen the population of the capital Port-au-Prince balloon out of control as farmers who cannot feed themselves move to the city in search of employment."[18] One dot connects with another. The 2010 earthquake that devastated the island and killed 230,000 cannot be viewed in isolation from the background of this urban shift. More people living in a concentrated area correlates to a higher death toll. The population of Port-au-Prince in 1980 was 700,592. The population in 2010 was 2,140,638; a threefold increase. The housing quality built for this newer, indigent population was grossly inadequate. An economic policy of trade liberalization imposed on the government triggered further collapse of their rural agricultural base, increased the urban population, and exposed more to the suffering of a natural disaster.

In recent years, a major drought developed into famine in the Horn of Africa and throughout 2011 and 2012, large numbers fled, particularly Somalis and Ethiopians, in search of assistance and protection. The majority went to Kenya or Yemen; some traveled to more distant countries like Egypt. And now, recent reports say failing rains, severe malnourishment, enduring conflict, and poor sanitation mean Somalia is again facing a humanitarian crisis, with 50,000 children "at death's door" and 2.9 million Somalis at risk of hunger [19].

Without exception, displaced persons mentioned (lack of) livelihood options as one of the main reasons for leaving their homes and going to Kenya, Egypt, or Yemen. Most respondents were pastoralists, small-scale farmers, and agro-pastoralists. The displaced persons were mostly concerned about limited rights to work and lack of livelihood options. Other concerns included shelter, food security, education, access to health care, security, including protection against gender-based violence, and durable solutions, such as resettlement [20].

The collapse of rural agricultural communities due to ecosystem dysfunction, local and global (such as climate change), forces population shifts into urban areas or to other areas where land can be farmed. In Kenya, the Mau Forest was one such place of refuge. However, in their attempt to save the forest's ecosystem and reduce deforestation, evictions were necessary. The Mau Forest is the source of at least 12 rivers feeding into 3 lakes within the East African region. "Evictions in 2018 and 2019 have targeted people who have settled on Maasai Mau, a block of Mau forest managed by the Narok county government that is held in trust under the Mau Trust Land. The Kenya Forest Service manages another 21 blocks." [21]

Environmental degradation and drought have severely disrupted rural livelihoods in Ethiopia, Iraq, Somalia, and Niger. Yet drought means much more than just displacement and the loss of a job for the thousands of affected pastoralists and farmers. In many cases drought means abandoning an ancestral way of life. People become displaced when their livelihoods reach a critical threshold, below which pastoralism or farming is unsustainable. "What is the tipping point [22]?

The Sankofa principle places a burden on those who have studied the past to be motivated into action to save the future. But as countries urbanize, the decision-makers for agriculture policy are less the ministries of agriculture who are connected to the farming situation and more to city planners, mayors, district councils, trader organizations, and public health professionals. These hold more power in crafting agriculture policy, keeping markets functioning, and overseeing food processing and safety systems. The approaches taken by the government to agricultural production are shaped by ideas of economic development, economic interests, the prescriptions and requirements of international agencies (such as the World Bank and the International Monetary Fund), and by regimes, local environmental conditions, and legacies of national and sub-national institutions among others [23].

Lessons To Guide Our Future

What are the lessons to learn from the past regarding agricultural policy and technology? The rural and urban landscapes where we each claim global citizenship have been poisoned, overworked, mistreated, drained of their inherent value, and then abandoned for dead. The flight to the cities is a consequence of failed ecosystems in the agricultural regions. Considering that more that 44% of the sub-Saharan employment is connected to agriculture, the contraction of agricultural populations to cities is a foreboding statistic. The conventional solution will be to mechanize more and focus on land productivity through agricultural technologies.

Appropriate agriculture strategies begin with healing the land and restoring biodiversity. It returns sovereignty to communities from the beginning to the end of the food chain, cutting their economic chains of debt, dependence, and discrimination. Appropriate agriculture enhances economic and biological circularity, regenerative and synergistic life systems, environmental stewardship, and sufficiency over greed. Technologies aligned with these values are appropriate. On the other hand, input intensive agricultural technologies are proven to destroy biodiversity and to weaken the sovereignty of farmers. Technologies that facilitate this agenda are deemed in-appropri ate. All technologists who, knowingly or unknowingly, support the advancement and implementation of such technologies are irresponsibly contributing to a food and water crisis that has already made itself manifest.

There are multiple cases in history that serve as evidence that ecological deficits exist around the globe. Their consequences have been disastrous for the human population. Community poverty is the consequence of multiple compounding historic factors and policy decisions. Unfortunately, those who have learned the lessons are rarely the decision-makers. They are either the victims or they are the studious observers who meticulously record the relationship between failed public policy, land law, corporate hegemony on a multinational scale, and the global poverty which ensues. Sankofa requires us to look back with the intention of learning lessons to guide the future.

The global decision-makers are rarely looking back and when they do, it is not to inform their future decisions. The 2021 COP26 (Conference of the Parties) in Glasgow, Scotland affirms this point. While there was consensus on the legitimacy of climate change and its global economic impact, the poor nations demanded the richer nations financially compensate them as an offset to their irreparable harsh realities. Their call fell on deaf ears and the motion was blocked. Some hope was found in the closing remarks by UN Secretary-General Antonio Guterres. He recognized what he called the "climate action army." Guterres acknowledged the power of activists to propel governments and companies beyond words and into action. He urged them: "Never give up. Never retreat. Keep pushing forward." [24]

Regenerative Agriculture

Regenerative agriculture (agroecology) is a sustainable approach to agriculture, the science of cultivating soil, producing crops, and raising livestock. A functioning agricultural system requires a thriving biodiversity. Biodiversity is the bedrock of productive agriculture. In addition to the provisioning services associated with crop and animal production, biodiversity contributes to ecosystem services that benefit agriculture and society more broadly. These include higher yield and overall production output through intercropping and agroforestry, regulation of pest and diseases, nutrient cycling though decomposition of organic matter, carbon sequestration, soil water retention, and pollination services. Crop diversity is one example of the benefits of increasing biodiversity. In addition to enabling the production of food across a wide spectrum of environmental conditions, crop diversity (especially fruits and vegetables) contributes to food security, a diversified diet and higher quality nutrition [25].

Permaculture and Biodiversity

We are symbiotic beings. By definition, by our ethics, and by the very laws of nature, in order for an ecological design to succeed, we must remove ourselves from the throne. We must make the conscious choice for humanity's needs to be on equal ground with the needs of other species. Fortunately, this is the ecological default. We just have to stop working against nature. We have to embrace the fact that, in order for our own species to survive, we have to build an active perpetuation of others into our dedicated daily life patterns.

According to Dr. Lynn Margulis, an evolutionary biologist who co-created the Gaia Theory and proved several of Darwin's assertions to be wrong, "evolution was/is dependent on collaboration and cooperation, NOT just competition!" [26]

Permaculture is an approach to land management and settlement design that adopts arrangements observed in flourishing natural ecosystems. It includes a set of twelve design principles derived using whole systems thinking. These principles are closely aligned to the appropriate technology movement and reinforce the importance of increasing biodiversity as the basis of regenerative agriculture. The permaculture principles are detailed below.

12 Principles of Permaculture

1. Observe and interact. Every site is unique and its ecological balance must be understood before actions are taken. The goal is to "dance with nature" not fight it.
2. Catch and store energy. When resources are abundant, it's smart to store some of them to use later. This would include solar or wind energy and rainwater harvesting.
3. Obtain a yield by letting nothing go to waste. Constantly look for opportunities to create synergies and collaborations with all facets of the land. This is where biodiversity thrives.
4. Apply self-regulation and accept feedback from Nature. Your land is an ecosystem and not all elements may yet be in balance. Be observant and respond to the messages by making adjustments.
5. Use and value renewable resources and services. Choose energy and materials that replenish themselves. Keep your soil rich and healthy by building up the microbes in your soil with compost, and choose biodegradable materials to suppress weeds.
6. Produce no waste. Repair and reuse. Protect your tools and equipment. Create closed loops — feed your food scraps to the chickens, they produce manure, the manure turns into compost and is added to the garden, the garden produces veggies to eat and more food scraps.
7. Design from patterns to details. The logistics of how we interact within the ecosystem must be observed and designed accordingly. Kitchen herbs and medicinal herbs should be in zones close to the home.
8. Integrate rather than segregate. Operate as a community rather than as an individual. Share resources with your neighbor and build connections between homesteads for community sufficiency.
9. Use small and slow solutions. Be patient and think long-term. Going slowly means we save on resources because we won't buy new things that end up wasted.
10. Use and value diversity. Growing a diversity of plants means there is always something for beneficial insects to eat, and

protects your garden from pests and disease. Having a good mix of animal life in and around the farm will always be advantageous.

11. Use edges and value the marginal. Herbs like nettle and dandelion can be foraged from the edges of footpaths or unused plots of land.
12. Creatively use and respond to change. Crop rotation is a way of creating change and resting or replenishing soil. This leads to a reduction and reduced risk of pest and weed infestations. Diverse root systems create greater distribution channels or biopores. As roots explore the soil profile for nutrients and water in diverse strata it results in a higher utilization of these resources.

Technologies and processes, such as permaculture, which embrace regenerative agriculture are appropriate. We now look at the role of appropriate agricultural technology for soil health, seed quality, and farm mechanization.

Soil Health

Improving soil health is considered an essential building block in the biodiversity equation. Soils harbor a large portion of global biodiversity, including microorganisms (e.g., bacteria), micro- (e.g., Nematoda), meso- (e.g., Collembola), and macrofauna (e.g., Oligochaeta). This high biodiversity plays critical roles driving multiple ecosystem functions and services, including climate regulation, nutrient cycling, and food production [27]. Accordingly, recent experimental and observational studies, based either on particular biomes (e.g., drylands) or local sites, have shown that soil biodiversity is of high importance for the maintenance of multi-functionality (i.e. the ability of ecosystems to simultaneously provide multiple ecosystem functions and services in terrestrial ecosystems.) [28]

Modern agriculture has resulted in the destruction of soil biodiversity and dramatically decreased soil health. The focus on increasing agricultural production to meet the rising urban populations have relied on intensification, concentration and specialization of agriculture, relying upon new technologies of agricultural chemicals, mechanization, and plant breeding (hybrids and GMO's). The undesirable consequences have reduced environmental quality and natural resources as well as community health and economic insecurity for the smallholder farmer.

Regenerative (sustainable) agriculture represents a way to heal and restore lost biodiversity. Once restored, sustainable agriculture refers to the ability of a farm to produce food indefinitely, without causing severe or irreversible damage to ecosystem health. As for the more general concept of "sustainability", three key issues are involved: environmental (the long-term effects of various practices on soil properties and processes essential for crop productivity), social, and economical. Soil biodiversity is essential for agriculture and food security and its proper management will aid agriculture being carried out using sustainable methods [29].

Seeds for the Sahel

One area of technology critical to agricultural success are seeds and seed quality. The seedball is an example of an appropriate innovation in how seeds can be spread effectively in regions where conditions are difficult. To improve seeding success rates and combat the Sahel's harsh conditions, the seedball is a tiny capsule that may revolutionize Sahelian cropping systems. The technology comes out of the Feed the Future Innovation Lab for Sorghum and Millet, led by Kansas State University and Ludger Herrmann, a researcher from the University of Hohenheim. The technology is part of a long-term collaborative research effort with farmer organization Fuma-Gaskiya in the Maradi region of Niger. Sahelian farmers often have limited space for cropping, very low incomes, and restricted access to inputs such as fertilizer and pesticides. This combination of limitations makes it difficult for farmers to subsist on their cropping operations, and nearly impossible to sell their crops for income.

The seedball is small enough to fit in the palm of your hand. The plants emerging from seedballs have higher drought tolerance and can survive longer dry spells than plants that are dry scattered as seeds. They increase labor efficiency and reduce the need to replant.

"These fields often fail to produce yield due to the fact that plants emerge after little rain that is followed by drought," Herrmann said. "The nutrient formulation of the seedballs should support early plant growth, root growth in particular, so that seedlings are able to exploit a larger soil volume." The long-term goal is to define additives so seedball seeds germinate with a defined minimum amount of rainfall. "We believe that seedballs can improve the local millet and sorghum cropping systems by reducing seed expenditures, increasing seedling survival, enhancing nutrient and water use, and finally by reducing cropping risk, increasing yield and thus income and resilience," Herrmann said [30].

The Role of Mechanization

Mechanization in agricultural crop production and livestock management is an important component of defining appropriate technology. Increasing agricultural power through mechanization means that more gets done in the limited time available. Where one has to be cautious is in the matter of scale. As small-scale farmers lost their land access to corporate agriculture, the size of a farm became exponentially enormous. Monoculture farming was in the driver's seat. The large investment required for mechanization kept medium-scale farmers in debt and enslaved to the land and machines 'they worked for'. In the U.S., the insatiable need for cheap labor evolved into the employment, legal and illegal, of migrant farm workers from south of the border. Similarly, when corporate agriculture migrated, it was to an abundant supply of cheap labor in almost every corner of the developing world.

Mechanization becomes dehumanizing and inappropriate when its scale regards humans as adjuncts to the machines in the pursuit of profit. When agricultural mechanization is applied at the proper scale, it can contribute significantly to the development of value chains and food systems. Mechanization has the potential to render post-harvest, processing, and marketing activities and functions more efficient, effective and environmentally friendly [31]. Farmers who have access to improved agricultural tools and powered technologies can shift from subsistence farming to more market-oriented farming, making the agricultural sector more attractive to rural youth. Sustainable mechanization supports the development of food supply chains through improved agricultural practices for increased production and enhanced food security. Opportunities for appropriate mechanization include land preparation, timely seeding and planting, weed control, pest management, precise fertilizer application, harvesting, preparation for storage, and value addition operations along the food supply chain in terms of on-farm processing, transport, and marketing.

Guidelines for appropriateness would not be limited to work efficiency but would encompass minimal disruption to ecological balance and biodiversity. The use of clean energy would be essential. One example is growing crops without mechanical seedbed preparation and with minimal soil disturbance since the harvest of the previous crop (no-till farming). Crop residues are retained either completely or to a suitable amount to guarantee the complete soil cover, and fertilizer and amendments are either broadcast on the soil surface or applied during seeding. Keeping the soil covered is a fundamental principle of sustainable agriculture leading to improvement of soil properties. It also increases soil biodiversity.

Information Access and Best Management Practices

Regenerative agriculture is rooted in endogenous knowledge at the regional level and evidence-based best practices at the global level. It is here that information technology has a critical role to play: making connections between the farmers who produce and the researchers who organize and analyze regional and global behaviors; connecting producers to the marketplace; and connecting farmer groups with common interests with each other.

The *Chakula Bora Digital Platform* [32] is one such digital space that hopes to give birth to a suite of disruptive innovations, initiatives, resources, and policies which lead the way to more effective agricultural development policies. The platform seeks to curate and share research on the various drivers of agri-food systems in Africa. It will examine how change affects these drivers and how these changes impact food availability, markets, and consumptions. The digital platform will foster linkages and collaborations among key stakeholders in the entire agricultural sector who share the same goals and objectives. It will empower a community of actors in the regenerative agriculture space who, working collectively, will bring about true food sovereignty, grounded in sustainable food systems.

Specific knowledge areas of the *Chakula Bora Digital Platform* include: organic methods of soil improvement and soil health, training farmers on best crop and animal husbandry practices, seed production and distribution, pest management and early disease detection, and sustainable production systems. Education and training is another functional area. It includes primary and secondary school training in agroecology and sustainability; training resources for rural extension service staff to plan, implement, and monitor their projects, in collaboration with their local communities and farmer groups.

Appropriateness in Agriculture

Regenerative agriculture is appropriate for its many social benefits: it strengthens communities, advances gender equity, and promotes health and wellness. Regenerative agriculture offers many technical strategies which rebuild and harmonize with the ecosystem. They include:

1. Mimic the natural forest systems. Adopt agro-ecological production models such as conservation agriculture, organic agriculture, regenerative agriculture, permaculture, etc.
2. Cultivate protein crops on grasslands. This supports carbon capture as well as animal feed.

3. Phase-Out plant protection products and their dependencies [33].
4. Diversify Crops. Crop diversification, whether through crop cohabitation or seasonal or multi-annual diversification within the same plots of land, is an invaluable strategy.
5. Diversify Seeds. Planting different seed varieties of the same crop within a plot increases biodiversity on that plot.
6. Restore natural habitats in intermediate areas. Widen the flower strips that border conventionally farmed fields.
7. Develop solutions for re-establishing biodiversity in intermediate spaces, such as roadsides [34].

Food Sovereignty Not Food Security

What role does the farmer play in a sustainable production system? How can we co-construct sustainable solutions with them, recognizing their critical role in our survival and the future of our planet?

Agriculture represents the sum of processes to sustain human life through food production. Within this domain of life sustenance, technologies have their role. From cradle-to-grave to rebirth, life feeds on and produces life. All seven nutrients needed to sustain life are found within life itself. Appropriate agricultural technologies aim to complement these natural processes. When these technologies introduce inorganic processes and synthetic chemicals into the process, they kill life and become inappropriate.

In the name of food security, the 'green revolution' has literally 'made a killing'. Ecosystems have been destroyed. Fertile land has been poisoned and overworked, producing less and less. The cause of global hunger includes the agricultural processes themselves. The solution advocated in the pursuit of food security (and short-term profits), has been to acquire more land, displace more people, and increase inorganic inputs. Food security, when viewed as a social welfare problem, is addressed by importing processed food back into the source communities.

Mahatma Gandhi said, "Earth provides enough to satisfy every man's needs, but not every man's greed." Stable ecosystems have been so maligned that biodiversity has been lost, farmland is no longer productive, forests have been castrated of their power, and a polite name has been assigned to our frequent droughts, typhoons, hurricanes, and melting ice caps ... Climate Change. Global warming (climate change) is the culmination of a greed that knows no end. The food chain is exasperated because of it; hence, the outcry for food sovereignty. The communities that produce our food are demanding a return of control of the agricultural processes that define our food chain.

The scale of hunger and malnutrition across the world today is the direct result of a global economy in which hundreds of millions of small farmers, fisherfolk, pastoralists, and indigenous people face ruin - because of the hijacking of the food system by large corporate agribusiness and food retailers. Decisions about what is produced, what is consumed, and who has access to food are defined by multinational corporations in control of the entire food chain and by trade rules by institutions such as the World Trade Organization.

A popular belief that the world's population has grown too large for us to produce enough food to feed everyone is completely false [35]. Large global agribusinesses are responsible for the overproduction of crops and the resulting vast amounts of food waste throughout the entire supply chain, from field to the supermarket shelf. Their greed has led to imbalances in our biosphere which has spiraled out of control. Food sovereignty is the pathway to restoring that control.

The Nyéléni Declaration for Food Sovereignty

Food is a fundamental human right. Appropriate technologies must stand behind the principles of food sovereignty, not food security. In the name of food sovereignty, technologies must be guided by the sustainable care and use of natural resources. This includes air, land, and water, but also seeds and livestock breeds. Appropriate technologies put people and the planet before profit.

Those who work the land must do it using sustainable methods that conserve biodiversity while deriving the first fruits of their labor. To deny them this right is inappropriate. Corporate control of the global food systems is inappropriate.

Food sovereignty aims to reduce hunger and feed the world's population with nutritious, healthy food, produced sustainably and sold at a fair price. The movement had its genesis in 1996 at the La Via Campesina, where millions of small-scale farmers gathered from around the world. The concept was introduced at the World Food Summit in Rome. By 2007, the six principles of food sovereignty were presented in Mali and the Nyéléni Declaration for Food Sovereignty was signed. The six principles of the Nyéléni Declaration are repeated here as they are essential guideposts to the development of appropriate technologies and agricultural policy.

1. *Food for people:* All individuals, people groups, and communities have the right to sufficient, healthy, and culturally appropriate food. Specifically mentioned are those who are already reeling from global warming's impact: all who are hungry, under occupation, in conflict zones, and marginalized. It rejects the proposition that food is just another commodity or component for international agribusiness to trade.
2. *Value food providers:* The contributions and rights of all people who cultivate, grow, harvest, and process food must be valued and respected. It rejects policies, actions and programs that undervalue them, threaten their livelihoods, and eliminate them.
3. *Localize food systems:* Bring providers and consumers closer together and put them at the center of decision-making on food issues; protect food providers from the dumping of food and food aid in local markets; protect consumers from poor quality and unhealthy food, inappropriate food aid and food tainted with genetically modified organisms; and resist governance structures, agreements and practices that depend on and promote unsustainable and inequitable international trade and give power to remote and unaccountable corporations.

4. *Localize control:* Place control over territory, land, grazing, water, seeds, livestock, and fish populations on local food providers and respect their rights. They can use and share them in socially and environmentally sustainable ways which conserve diversity; it recognizes that local territories often cross geopolitical borders and ensures the right of local communities to inhabit and use their territories; it promotes positive interaction between food providers in different regions and territories and from different sectors that helps resolve internal conflicts or conflicts with local and national authorities; and rejects the privatization of natural resources through laws, commercial contracts, and intellectual property rights regimes.
5. *Build on knowledge and skills:* Develop and enhance the skills and local knowledge of food providers and their local organizations that conserve, develop, and manage localized food production and harvesting systems, developing appropriate research systems to support this and passing on this wisdom to future generations; and rejects technologies that undermine, threaten or contaminate these, e.g. genetic engineering.
6. *Work with nature:* Understand and use the contributions of nature in diverse, low external input agro-ecological production and harvesting methods that maximize the contribution of ec systems and improve resilience and adaptation, especially in the face of climate change. It seeks to heal the planet so that the planet may heal us; and, rejects methods that harm beneficial ecosystem functions, that depend on energy intensive monocultures and livestock factories, destructive fishing practices and other industrialized production methods, which damage the environment and contribute to global warming [36].

Food Sovereignty's Priorities

One of the top priorities of the food sovereignty movement is control over seeds. Controlling seeds impacts the entire food chain. Peasants' ability to access seeds, their collective rights to use, produce, preserve, exchange and sell them, as well as the autonomy of peasant seed systems, are major issues for peasant organizations and civil society activists. Seeds are central for agroecology.

There is no agroecology without a guaranteed farmers' seeds system. Food sovereignty respects and protects farmers' rights to use, save, exchange, and sell seeds, putting the control of biodiversity and knowledge back in the hands of peasants. Agricultural policies must value local knowledge and give opportunities to share peasants' knowledge.

The food sovereignty movement is also concerned with declining animal biodiversity prompted by inappropriate interventions. The emergence of a zoonotic pathogen in humans, such as SARS-CoV-2, was not unpredicted. 75% of the infectious diseases affecting humans in the 21st Century are of zoonotic origin. The combination of genetically similar animals in unhealthy production environments (with their attendant over-use of antimicrobials) is a global pandemic in the making. Governments should move rapidly to remove barriers to small-scale livestock farmers and pastoralists, and stop subsidizing industrial livestock agriculture that is destroying ecosystems from the genes up [37].

Appropriate agricultural technologies embrace the principles of food sovereignty. Increasingly, the domain of these technologies include the field of genetics and pollution-induced mutations. Technologies must restore the ecological imbalances created by industrial and destructive fishing practices, climate change, water contamination caused by mining, the proliferation of invasive species, large-scale infrastructure development, violence and persecution, water grabbing, privatization, and exclusion of the natural resources on which they depend.

Conclusions and Policy Recommendations

Food and water access are essential for human life. The development of a manifesto for appropriate technology must regard agriculture and its food chain highly. Appropriate technology in agriculture must understand the past and extract its hard lessons to define solutions. It must also respect the principles of local control as expressed by food sovereignty. It must seek more than to be nature friendly. It must work to heal nature, in light of humanity's past sins. It must observe and practice imitating nature's simple but effective processes. Throughout the food chain, from farm to fork, regenerative agriculture and nature-inspired systems must be evident.

Who Makes the Decisions?

There are 17 sustainable development goals defined by the United Nations. They are presented in Table 1 below. All can be substantially addressed using the principles of sustainable agriculture as presented in this chapter. The correlations are well-documented. The obvious question is why is this not happening? The message of Sankofa is to look back to move forward. Those who are looking back are frequently not the decision-makers. Those who have realized the lessons to be learned, who have recorded and correlated the data and published their conclusions, are not the ones guiding public policy. They are not the leaders of nation-states who say "yeah or nay" to the "captains of industry." . Decision-makers are the ones who must be incentivized to think sustainably.

One overwhelming challenge to implementation of sustainable (appropriate) agriculture may be called "linguistic confusion" where the public understanding of the global crisis is strategically softened with words. Language has been strategically manipulated through media to confuse what is good and what is bad. What is bad for the planet or the economy has been rephrased into words that are "not so bad." For example, "green revolution" sounds like something all proponents of food sufficiency must embrace; "food insecurity" sounds like a very adequate description of the millennial global challenge of depleted and depleting natural resources; "climate change" allows me to sleep at night as compared to "global warming", "agricultural productivity" suggests that for a given plot of land, one can invest inputs and use processes which maximize the return on those investments. The appeal of softer language combined with a sprinkling of money has helped global life-threatening activities continue without panic, from the average well-meaning citizen to the halls of academic research.

Table 1: Sustainable Agriculture and The Sdgs

Relative correlation between the practices of sustainable agriculture and achieving the 17 United Nations Sustainable Development Goals

Moderate to High Cause-Effect Correlation / C. Low to No Cause-Effect Correlation

U.N Sustainable Development Goals	Restore Ecological Balance	Regenerative Agriculture Instead of Monocrops	Embrace Food Sovereignty
1. Reduce Poverty	A	A	A
2. Reduce Hunger	A	A	A
3. Increase Health & Well-Being	A	A	A
4. Accessible Quality Education	C	C	A
5. Gender Equity	C	C	A
6. Clean Water Access & Proper Sanitation	A	A	A
7. Renewable and Clean Energy	A	C	C
8. Meaningful Work and Balanced Economic Growth	C	A	A
9. Increased Innovation and Stronger Community Infrastructure	C	C	A
10. Reduced Social Inequalities	A	C	A
11. Sustainable Cities and Communities	A	A	A
12. Healthy Consumption and Production Behaviors	A	A	A
13. Corrective Action to Global Warming	A	A	A
14. Restored Water Ecosystems	A	A	C
15. Restored Land Ecosystems	A	A	C
16. Peace, Justice, and Strong Institutions	A	C	A
17. Partnerships for the Goals	C	C	A

Recently, the political buzz has been about the "Green Economy." The term certainly represents a push for jobs and policies that support a healthy planet. Yet, if big money is to be allocated, it is certain we will find a number of "rotten eggs within that carton." Venezuelan professor Edgardo Lander says that the UNEP's report, Towards a Green Economy, while well-intentioned "ignores the fact that the capacity of existing political systems to establish regulations and restrictions to the free opera tion of the markets – even when a large majority of the population call for them – is seriously limited by the political and financial power of the corporations."[38]

The Role and Responsibilities of Higher Education

What role can higher education institutions play in advancing sustainable agriculture?

1. They are responsible for sounding the alarm! Our food chain has been hijacked from smallholder farmers and transformed into a global machine for corporate profit which has no boundaries. Academic curriculums must extend across multiple disciplines. Economic strength is linked to ecological health and to public health. The technological solutions applied to agricultural productivity must be human-centered and recognize the importance of biodiversity. The voice of civil society must be strengthened.
2. Don't compromise the laudable ethics embedded in the mission of the university when in the pursuit of funding.
3. Be active in regenerative agriculture research, collaborating with local farmer cooperatives.
4. Be the legitimate source of research that shapes national policy on climate change and its cause-effect relationship with food, water, health, and community resilience.

NGOs, International Aid Agencies, and Nonprofits

1. Non-governmental, nonprofits, and aid agencies serve a vital role in lifting up and protecting the most marginalized and the global environment we all share. It is important that they work together since their missions and visions are intricately intertwined.

2. Protocols embracing handouts to the impoverished and that create dependency must be replaced with an emphasis on sovereignty, creating circular local economies at scale, resilience, and embracing entrepreneurship.

The Role and Responsibilities of Government

1. Public policy is a powerful agent for economic strength and resilience. Leaders properly informed and educated can make sustainable decisions, benefiting generations to come.
2. In a world where nation-states are easy prey for multinational corporations, government leaders at every level must be accountable to the people they represent and to their oath of office. Laws are in place but are often not enforced with accountable structures.
3. "Teach them when they are young." Programs in primary and secondary schools that connect the dots between nature's biodiversity, organic farming, and healthy food should be widely promoted. The agricultural sector is collapsing but it is the economic sector necessary for humanity to survive.
4. Promote food sovereignty along with building strong institutions for civil society.
5. Strengthen the national ministry of agriculture with a mission to rebuild from the small farm up. Provide regional agricultural extension offices empowered to empower based on regenerative agriculture. The pathway towards addressing the social ills in our urban populations begins with reinforcing the rural areas. The high status associated with growing urban populations cannot negate the importance of having a strong rural base. Small is still very beautiful.

Bibliography

[1] MacDonald, J. M. (2014, March 4). Family Farming in the United States. U.S. DEPARTMENT OF AGRICULTURE, Economic Research Service. https://www.ers.usda.gov/amber-waves/2014/march/family-farming-in-the-united-states/

[2] MacDonald, James M., Penni Korb, and Robert A. Hoppe. Farm Size and the Organization of U.S. Crop Farming, ERR 152. U.S. Department of Agriculture, Economic Research Service, August 2013

[3] GMWATCH. (2012, October 22). *NEW Report Critical of Gates/Rockefeller "Green Revolution for Africa" / DuPont and CIMMYT (21/10/2006).* https://www.gmwatch.org/en/news/archive/2006/5123-new-report-critical-of-gates-rockefeller-qgreen-revolution-for-africaq-dupont-and-cimmyt-21102006

[4] GMWATCH. (2012, October 22)

[5] GMWATCH. (2012, October 22)

[6] Anwar, M. A. (2017, April 24). *White people in South Africa still hold the lion's share of all forms of capital.* The Conversation. https://theconversation.com/white-people-in-south-africa-still-hold-the-lions-share-of-all-formsof-capital-75510

[7] Clark, N. L. (2014). *Structured Inequality: Historical Real ities of the Post-Apartheid Economy.* Escholarship. https://escholarship.org/content/qt7wb1g2wp qt7wb1g2wp_noSplash_99967523a78cb6518a8aedc7cb10c716.pdf?t=ngmi8e

[8] Clark, N. L. (2014)

[9] de Castro Pena, J. C., Goulart, F., Fernandes, G. W., Hoff mann, D., Leite, F. S. F., dos Santos, N. B., Soares-Filho, B., Sobral-Souza, T., Vancine, M. H., & Rodrigues, M. (2017, July 1). *Impacts of mining activities on the potential geo graphic distribution of eastern Brazil mountaintop endemic species.* ScienceDirect. https://www.sciencedirect.com/science/article/pii/S167900731630130X#bib0180

[10] Abbott, E. *"Haiti's Revolution Fueled the Rise of Big Sugar in Cuba and Louisiana".* The George Washington Univer sity, History News Network. https://historynewsnetwork.org/article/125294

[11] Food and Agriculture Organization of the United Nations, Sustainable Development Office. (1999) "Population and Deforestation".

[12] Byrd, Brandon. Haiti: The Black Republic , p28 (2019)

[13] REGAN, J. (2003, August 5). *Forest land in Haiti fading fast.* Latin American Studies. http://www.latinamericanstudies.org/haiti/haiti-deforestation.htm

[14] Coulibaly, B., Sagoe, G., & Shixiang, L. (2021, March 4). *Towards poverty alleviation in developing countries: An empirical study of the impact of land tenure reforms in Kati, Mali.* Plos One. https://journals.plos.org/plosone/article?id=10.1371/journal.pone.0246502#pone.0246502.ref004

[15] Coulibaly, B. et al. (2021, March 4).

[16] Coulibaly, B. et al. (2021, March 4).

[17] Coulibaly, B. et al. (2021, March 4).

[18] Jerving, S. (2020, September 9). *5 ways urbanization is changing agriculture in Africa.* Devex. https://www.devex.com/news/5-ways-urbanization-is-changing-agriculture-in-africa-98045

[19] United Nations Department of Economic and Social Affairs. (2018, May 16). *68% of the world population projected to live in urban areas by 2050, says UN.* https://www.un.org/development/desa/en/news/population/2018-revision-of-world-urbanization-prospects.html

[20] Doyle, M. (5 October 2010). *US urged to stop Haiti rice subsidies.* BBC International News.

[21] Smith, C. (2014, May 27). *Understanding Disaster-related Displacement from the Horn of Africa.* Our World. https://ourworld.unu.edu/en/understanding-disaster-related-displacement-from-the-horn-of-africa

[22] Smith, C. (2014, May 27)

[23] Human Rights Watch. (2020, October 28). *Kenya: Mau Forest Evictees' Plight Intensifies.* https://www.hrw.org/news/2020/07/23/kenya-mau-forest-evictees-plight-intensifies

[24] Internal Displacement Monitoring Centre. (2020, June 11). *Beyond drought: Adding life to the numbers.* https://www.internal-displacement.org/features/beyond-drought-niger-ethiopia-somalia-iraq-displacement

[25] Lencucha, R., Pal, N.E., Appau, A. *et al.* (2020) *Govern ment policy and agricultural production: A scoping review to inform research and policy on healthy agricultural*

commodities. Global Health 16, 11 (2020). https://doi.org/10.1186/s12992-020-0542-2.

[26] Hill, A. C., CFR Expert, & Babin, M. (2021, November 15). *What COP26 Did and Didn't Accomplish.* Council on Foreign Relations. https://www.cfr.org/in-brief/cop26-climate-outcomes-successes-failures-glasgow

[27] Chappell, M. J., Wittman, H., Bacon, C. M., Ferguson, B. G., Barrios, L. G., Barrios, R. G., Jaffee, D., Lima, J., Méndez, V. E., Morales, H., Soto-Pinto, L., Vandermeer, J., & Pefecto, I. (2013, November 6). *Food sovereignty: an alternative paradigm for poverty reduction and biodiversity conservation in Latin America.* US National Library of Medicine. National Institutes of Health. https://www.ncbi.nlm.nih.gov/pmc/articles/PMC3869480/

[28] Flores, H. J. (2022, January 3). *Biodiversity is the Staff of Life.* Freepermaculture. https://www.freepermaculture.com/biodiversity/

[29] Guerra, C.A., Heintz-Buschart, A., Sikorski, J. *et al.* Blind spots in global soil biodiversity and ecosystem function research. *Nat Commun* 11, 3870 (2020). https://doi.org/10.1038/s41467-020-17688-2

[30] Guerra, C.A. *et al*, 2020

[31] Guerra, C.A. *et al*, 2020

[32] United Nations, Office for the Coordination of Humanitarian Affairs. (2015, October 5). *Giving Seeds an Added Boost to Survive Harsh Sahelian Climates* [Press release]. https://reliefweb.int/report/world/giving-seeds-added-boost-survive-harsh-sahelian-climates

[33] Food and Agriculture Organization of the United States (FAO). (2022). *Sustainable Agricultural Mechanization.* https://www.fao.org/sustainable-agricultural-mechanization/overview/why-mechanization-is-important/en/

[34] Stephenson, B. (2021). Chakula Bora Digital Platform is a web-based platform for collaboration and knowledge exchange in regenerative agriculture by Passion of Hope International.

[35] IDDRI. (2018) *An Agroecological Europe in 2050: multi functional agriculture for healthy eating, Findings from the Ten Years For Agroecology* (TYFA)

[36] Planet A. (2020). *Food Sovereignty and the Environment: Leverage to Reconcile and Guarantee Healthy, Sufficient Food that Respects Biodiversity*. Planet A Initiative. https://www.planet-a-initiative.com/food-sovereignty-and-the-environment-leverage-to-reconcile-and-guarantee-healthy-sufficient-food-that-respects-biodiversity/?lang=en
[37] War on Want. (2021, April 16). *Food Sovereignty: The Struggle for a Fair Global Food System*. https://waronwant.org/news-analysis/food-sovereignty-struggle-fair-global-food-system
[38] War on Want. (2021, April 16). *Food Sovereignty: The Struggle for a Fair Global Food System*. https://waronwant.org/news-analysis/food-sovereignty-struggle-fair-global-food-system
[39] Lawler, Odette K, et al. "The COVID-19 Pandemic Is Intricately Linked to Biodiversity Loss and Ecosystem Health." *The Lancet. Planetary Health*, vol. 5, no. 11, 1 Nov. 2021, pp. e840–e850, www.ncbi.nlm.nih.gov/pmc/articles/PMC8580505/, 10.1016/S2542-5196(21)00258-8.
[40] Lander, E. (2019, July 15). *The Green Economy: The wolf in sheep's clothing*. Transnational Institute. https://www.tni.org/en/publication/the-green-economy-the-wolf-in-sheep's-clothing

Chapter 9

Ethics and Appropriate Technology

Charles Verharen, John Tharakan, Flordeliz Bugarin, George Middendorf, David Schwartzman, Peter Schwartzman, Gada Kadoda, Bekele Gutema, and Enrico Wensing

Abstract

This chapter proposes an ethical foundation for appropriate technology practitioners based on the evolution of ethical systems developed in cultures in Africa and Eurasia. The authors are an international team of academics, activists and practitioners committed to an ethical approach to technology engagement. The proposed ethical model is congruent with the United Nations declarations of universal human rights. The model was developed to guide deployment of appropriate technology in the 21st century. Contemporary technology now enables humans to execute the destruction of life on earth through catastrophic climate change, weapons of mass destruction, the sixth mass extinction, global terrorism through genetic manipulation of viruses, or pandemics caused by viruses such as Covid-19. At the same time an appropriate use of contemporary technology can reduce energy and other resource scarcity, reduce the burden of human labor that can be executed by machines, and make possible universal education through post-secondary levels in both the Global North and South. The ethics section of the INAT Manifesto describes a philosophy that has the potential to guarantee life's future on earth.

Introduction

The inspiration for this chapter springs from the evolution of ethical systems over the past five thousand years. Cultures in ancient Africa, India and China developed the first formal ethical systems that focused on community harmony and control of the attention through meditation (Verharen *et al.* 2021). Greek cultures concentrated on reasoning dedicated to the prediction and control of experience. Christian cultures attempted to create a universal human community.

In more recent times, European cultures, bolstered by the advanced technologies of the industrial revolution, proposed a universal community grounded on maximizing the quantum of pleasure in the world. Darwin reminded his readers that survival is the precondition for speculating about ethics. German philosophers like Kant, Hegel, Marx and Nietzsche proposed freedom as the grounding ethical principle.

The proposed ethics model includes all the basic ethical principles developed in these diverse cultures as indispensable for ethics executed through appropriate technology in the 21st century. Contemporary technology now enables humans to execute the destruction of life on earth through catastrophic climate change, weapons of mass destruction, the sixth mass extinction, global terrorism through genetic manipulation of viruses, or pandemics caused by viruses such as Covid-19. At the same time an appropriate use of contemporary technology can reduce energy, food, water and other resource scarcity, reduce the burden of human labor, and make possible universal education. This chapter describes a philosophy that has the potential to guarantee both life's future on earth and more equitable lives for all nations.

Historical Review of Ethical Systems

There are at least two kinds of revolutions in ethics: one focused on defining what is good and the other on defining populations covered by ethical systems. Thinkers across Africa and Eurasia outlined the first kind of revolution over two thousand years ago.

While those same thinkers sketched proposals for the second kind of revolution at the same time, their proposals are only beginning to be acted on in recent times.

The term ethics has acquired the sense of a field distinct from and independent of morals. For the purposes of this essay, the term morals refers to behavior that is customary or acceptable in a given society (Haws 2004). Ethics includes the study of morals along with the study of value itself. What is valuable is what is desired or, more strictly, what is desirable given some set of fundamental assumptions. At its most basic level, ethics considers appropriate mechanisms for choosing values or principles to guide our lives.

A brief review of seven ethical systems that have *nominally* guided the lives of billions of humans over the centuries displays the remarkable diversity of ethics across time and culture. The catastrophic threats to life's future on earth demonstrate the need to create diversity in ethics in the present moment.

Ethics arise as an evolutionary process emerging through mythology, religion, philosophy, science and technology. In the most ancient writings on ethics, mythology, religion and philosophy are inextricably linked. The ancient Egyptians' primordial ethical principle some 5000 years ago was *Maat,* translated as harmony, peace, justice, order.

Anticipating both Thales and Heraclitus, the Egyptians believed that the universe emerged from a chaotic kind of watery mass called *Nun* to become the force of life, the sun, designated as *Ra*. Through an anthropomorphic act, *Ra* creates four elements akin to earth, air, fire and water. These elements in turn create human-like gods who are responsible for life on earth. The Egyptians invented a metaphysical system to amplify their ethical system's power. Immortality is granted only to those who religiously promote community harmony throughout their lives. Virtually all religions over the past five millennia have adopted elements of this system (Hornung 1990/1982, 1999/1995, 2001/1999; Assmann 1996; Faulkner 2005/1972; Verharen et al. 2014a).

Over two thousand years ago, Hindu and Buddhist sages commanded that we control our attention through meditation to gain eternal life freed from reincarnation's suffering. Plato and Aristotle focused their attention on reason's ability to discover or create patterns that can help us predict and control experience. Nonetheless, neither escaped religion's force as Plato inserted Hindu reincarnation into Book X of the *Republic* and Aristotle deployed an anthropomorphic "thinking god" as the universe's prime mover (McEvilley 2002). Two thousand years ago, Christ enjoined humans to form a single group bound together by universal and unconditional love. Jeremy Bentham and John Stuart Mill, inspired by technology's growing power, commanded humanity to universalize one of our most fundamental drives, pleasure (Verharen et al. 2013).

Moved by the science and technology of their times, Hegel and Nietzsche turned to nature and the universe itself for their philosophical inspiration. Prefacing Darwin, Hegel pronounced spirit or freedom as the "infinite impulse of activity to alter its forms" (*Philosophy of History* 1956/1837, p. 206). Contemplating Darwin's focus on survival through natural selection, Nietzsche commanded that all wrangling over philosophical injunctions for life's direction must be tested against a single incontrovertible value: Will experiments conducted in accord with these commands yield the human growth that best guarantees life's future?

Like Mill (2014/1859), Nietzsche envisions philosophy as an experimental 'science.' Thinking that a wiser humanity in a distant future will create an empirical science with the power to command life's future, Nietzsche contends that our present time is an "interregnum" where "it is best ... to found little experimental nations. We are experiments: let us also want to be such!" (*Dawn*, Book V [453], p. 457).

The history of philosophy—the true philosophy that creates powerful new "experiments" to ensure that life survives and flourishes—is the history of preposterous proposals that come to be regarded as common sense by the billions that *nominally* profess to be guided by their commands. Individually these experiments have guided their cultures for centuries—in the ancient Egyptian and Hindu cases for over three millennia.

The ethics model's hypothesis is that each discovered a value and a command indispensable for guaranteeing life's future: Control your attention through meditation (Hindus, Buddhists, Daoists). Live for school as the exercise of reason (Socrates, Plato, Aristotle). Live for love as universal and unconditional to create a universal human community (Christ). For pleasure as universalized (Bentham, Mill). To survive (Darwin). To be free (Kant, Hegel, Marx, Nietzsche). Or the very first written command emerging from ancient Egyptian ethics: Live to harmonize these invariably conflicting values and commands (Verharen *et al.* 2014b).

However, with the increased power of science and technology, these approaches have resulted in contemporary existential crises: catastrophic climate change, the sixth mass extinction, weapons of mass destruction, state or group terrorism and global pandemics (Gardiner 2011; Kolbert 2014, 2021; Schell 2000; Jamieson 2014). Of the approaches outlined above, one value has become clearly pre-eminent—survival. Yet our path to that goal is only possible through the wise exercise and collective impact of the other commanding values humans have created since writing's invention.

Calling this collective ethics "survival ethics" as embodied in the proposed ethics model puts too much stress on that single value. Extraordinary figures like Socrates, Christ, Gandhi, Biko, Malcolm X and King favored truth and love over survival. However, our survival's being at risk is the engine for promoting diversity in ethics.

The only secure path to survival is through diversification of values, through collectivization of multiple ethical frames. Diversity in values is the engine of diversity in human behavior. The rationale for that diversity emerges from Darwin's (1936/1871) four postulations for the survival and flourishing of communities. (Richardson 2004; Desmond & Moore 2009; Johnson 2010; Ridley 2016).

First, communities should aspire to increase group size. Given appropriate circumstances, larger groups have better chances of survival and flourishing. Humans instinctively increase their group sizes from families to tribes to tribal nations to nation states to empires. Unchecked, that instinct is catastrophic as the continuing contest of empires in the 20th and 21st centuries demonstrates. Philosophers and religious figures like the Buddha, Christ, Hegel and Marx projected a glorious future in which all humans would be united in a single community.

Second, groups should aspire to bond members so tightly together that they are willing to die for one another. Nations imagine that citizens owe their lives to the nation itself. Virtually all nations capitalize on this maxim by requiring compulsory military conscription in times of war, as well as sacrifices by citizenry. And many nations continue to insist that citizens who violate foundational community norms are subject to the death penalty.

Third, groups should ensure to the best of their ability that every member has an excellent life. The quality of a group's life is an aggregate of the quality of its members' lives. "All for one, one for all" is not an idle bromide. Biologist Wilson (2012, 2014) argues that, under favorable conditions, humans like nineteen other species are "eusocial" or genetically primed to sacrifice self-interest for group interest. Psychologist Pinker (2011, 2013, 2018) claims that humans are constantly seeking the "better angels of our nature" as science and technology give us more power to control our circumstances. Presumably, this results in better living conditions.

Fourth, groups should strive for maximum genetic, linguistic and other cultural diversity to the limits of their capability. This maxim flows from the principles of natural selection. We have no idea what nature will throw our way. Our only guarantee of humanity's future is to create as much diversity as we can command. Diversity in values is key to creation of all possible diversity. The constraint on the creation of values is their ability to guarantee life's future on earth.

The Proposed Ethics Model's Elements

The primary instrument of the proposed ethics model is a checklist that enumerates the basic conditions for survival—a tightly bonded group, clean air, temperature control through clothing and shelter, potable water, nutritious food, basic healthcare and education. The optimum guarantee of survival is through pursuit of the conditions for flourishing—rationality through continuing education, heightened community bonding, intelligent use of pleasure as a mechanism for guiding behavior, freedom defined as the potential for creativity, and introspection defined as the rational control of the attention, expansion of group size, and a constellation of these conditions (Verharen et al. 2011).

Classical African, Asian and European philosophers have designated each one of these conditions as a *summum bonum* for humanity. The proposed ethics model regards all seven conditions as indispensable to flourishing. The relative importance of each condition is a matter for decision in the context of local cultures. Maximum cultural variation is a precondition for the survival of the species across more than a small, limited environmental space (Locke 1989).

The ethics model entails a global network of research universities working together with government agencies, non-government agencies, small and medium enterprises, community-based organizations, concerned citizen-activists, and most importantly the vulnerable populations themselves. Together, they can drive towards the ethical, sustainable development of communities at risk. The ethics model embodies an emerging network of international centers dedicated to sustainable development grounded in social justice.

The objective of the network is to demonstrate that ethical development can achieve sustainable results through collaborative multi-disciplinary efforts focused on the implementation and evaluation of appropriate technology development projects. The model provides the flexibility to address challenges arising from the changing nature of both local and international development landscapes. It also speaks to how global trends in environment, culture, demographics, urbanization, economics, communication connectivity, healthcare, science and technology can be successfully incorporated.

One issue in development is the difficulty in isolating causal factors that affect the success of projects. In this regard, the proposed ethics model offers a checklist program to determine what works and what does not work. The use of checklists is critical to the design, implementation and assessment of development projects.

Both survival ethics theory and the partnership model conform to goals sanctioned by United Nations (United Nations 1948) Universal Declarations of Human Rights and the Millennium Development Goals (United Nations 2000). They also recognize the centrality of a community's right to determine its own survival strategy and to maintain its own identity and culture—both tangible and intangible (Singer 2011; Silversmith and Ruggles 2004; Singer and Stavenhagen 2004; Sen 2000).

The model's ethics teams would form themselves around a constellation of research questions. For the technical aspects of such a project, geologists, biologists, archaeologists, environmental anthropologists, chemists, engineers and technologists all have roles in decision and evaluation processes. For the social science aspects, economists, political scientists, psychologists and cultural anthropologists join the conversation. For the humanistic aspects, historians, philosophers, linguists, classicists and artists make their contributions.

In the interest of increased objectivity and improved outcomes, NGOs (Non-Governmental Organizations) add their expertise. Pursuing sustainability for the project includes participation of SMEs (Small and Medium Enterprises) that advise community members on forming a corporation for the maintenance of the well and for replacement wells in the long run. Most important is the participation of community members, both as individuals and as members of CBOs (Community Based Organizations).

The Key Elements of the Proposed Ethics Model Consist of the following:

- A core of interdisciplinary faculty trained in social and ethical concept and component inclusion in science and engineering education (Tharakan 2006).
- Faculty who recognize the potential of their students to promote community capacity-building (Tharakan *et al.* 2008).
- Networking among NGOs, SMEs, CBOs, community member partners, target organizations and individuals throughout the African Diaspora, and governmental agencies engaged in specific development projects—a synergistic outcome of faculty exchange and mentoring (Fortunak and King 2010).
- Revisions and inclusion of service-learning courses in technology curricula. Curricular offerings focus on: (1) ethics and the philosophy of technology and engineering; (2) appropriate technology for developing communities with focus on water, food security, sanitation, and environment; (3) alternative and renewable energy solutions in developing communities; (4) pharmaceutical manufacturing; (5) cultural and historical context, facilitation, and working with community members; (6) environmental, archaeological, and heritage conservation; (7) data collection, evaluation, assessment, and monitoring (Tharakan 2006, 2011, 2012; (8) and socially relevant computing (Kadoda 2012).
- A new generation of students trained in socially and ethically responsible engineering and social development. Students gain real world experience in development through mandatory curricular engineering service-learning projects (Tharakan 2012).
- Increased focus on development of curricular experiences, exposure and training in skills needed for careers in ecological and environmental areas (Cardelús & Middendorf 2013; Klemow *et al.* 2019; Middendorf *et al.* 2020).
- All teams learn from and work in tandem with the communities they hope to help (Fortunak & King 2010). As a pragmatic

approach to solving problems for populations in need, the proposed ethics model benefits university faculty and students as well as community members. "Out of the classroom and out in the field" community-based learning provides immediate research-guided action on local and regional issues of the day. Community-based learning has the potential to democratize education from primary through post-secondary levels.

The model teaches both students and faculty to deal with real-world issues in a real-world context (Balazs & Morello-Frosch 2013). The model incorporates community-based projects that are designed, developed and implemented with communities in partnership with NGOs, SMEs, CBOs for all phases .

The model's ethics teams have the expertise to pinpoint causes of inefficient subsistence strategies, document environmental features, plot landscape transformations, identify pollutants, predict impacts, and devise effective mitigation policies. They have the equipment to conduct archaeological excavations, soil and bone chemistry analysis, farming systems analysis, water quality assessments, tests for food-borne illnesses caused by natural resources, and investigations of air quality.

Environmental education and outreach are an integral component of all projects. They train community members and students, consult with NGOs and SMEs, and raise awareness of environmental control technologies. Standard Operating Procedures are used to manage key personnel who handle waste products or engage in practices that impact the environment. Partners and communities are educated about potential dangers and contaminants and made aware of the benefits of a clean and safe environment. (Verharen *et al.* 2013) The proposed ethics models sustainable development grounded in a collaborative, collective, multi-disciplinary framework. It leverages the expertise of academics, SMEs, and NGOs, while it relies on the energies and talents of young students. Most importantly, it is community-driven and recognizes the importance of empowering communities that are able to maintain their own solutions.

The Proposed Model as Applied Ethics with Working Examples

The model ethical platform is both theoretical and practical. The ethical goals are survival and flourishing for humans united into a single community bonded together with other life forms and the inorganic environment through the instruments of appropriate technology. The model's particular task is to develop research and implementation of those instruments to address the gaps between the rich and the poor whether in the global North and South.

The most urgent need is technologies that reduce resource and labor scarcity and increase access to university-level education. Confronting the current existential crises requires virtually every human's contribution. The model would target global movements such as ethical ecovillages that address extreme rural and urban poverty throughout the world. These movements must ensure that communities are able to flourish within the context of their own cultural traditions. They must to the degree possible guarantee the communities' self-sufficiency and autonomy through solarization and agroecology. And they must inculcate a sense of responsibility for neighboring communities in need of assistance. Ethical ecovillages will help those communities solarize their own villages, ensure access to basic needs for survival such as clean air, potable water and nourishing food, and promote access to education that moves the communities beyond survival to flourishing.

Contemporary forms of appropriate technology have the potential to reduce energy scarcity, promote recycling of other resources from "cradle to cradle," reduce the need for humans to perform labor that does not require creative human consciousness through AI, robotics and other machines; and deploy the increased leisure that machines permit in the service of universal education to the post-secondary level. Two examples of ecovillages in Africa and India illustrate the model's transition from theoretical to applied ethics.

Immersed in the study of European and African ethics, Professor Godfrey Nzamujo, a Nigerian Dominican priest, developed a model of rural ecovillages that embodies the proposed ethics model's principles (Nzamujo 2002). Three ethical principles guide the development of the Songhaï ecovillages. First, the villages must be autochthonous, rooted in the historical traditions of local cultures (Vodouhe & Zoundji, 2013). Indigenous knowledge forms the foundation of the village (Tharakan, 2017). Humanity's collective knowledge in the humanities, social sciences and STEM disciplines adds the superstructure. The complementarity of the old and new must be worked out to guarantee the survival and well-being of the community members. Communities that resist relevant changes to their long-standing traditions exercise autonomy at risk of disrupting the natural rhythms of their environment. Nzamujo's second ethical principle is that the village must be autonomous. Self-governance and independence depend on extensive knowledge.

Nzamujo envisions the ecovillage as a "knowledge enterprise," facilitated by information communication technology (ICT). With local knowledge and ICT access to the global pool of information, the village builds the foundation for self-reliance in producing the necessities of a flourishing life: energy, housing, clean air and water, food, healthcare and the advanced education that makes such a life possible (de Luca *et al.* 2013).

Nzamujo distinguishes between autonomy and independence (Nzamujo, personal communication, 15/05/2020). His version of autonomy stresses inter-dependence over independence. The ecovillages form a network of co-operation with one another and with neighboring communities. Communities that dismiss the principle of inter-dependence introduce technologies that deplete resources and devastate their environment.

Nzamujo's third ethical maxim is that the village must be authentic in two senses: ethical and practical. Authenticity has three distinct aspects. First, the village must be dedicated to the well-being of its residents and those of the networked ecovillages as well as the residents of neighboring communities. Second, the foundation of the village must be "authentic technologies…that generate only positive effects/benefits to the producers, consumers and the environment at the same time" (Nzamujo, Personal Communication, 15/05/2020). Such technologies enhance the flourishing of the villages and their neighbors.

Authentic technologies are defined as those "aligned with the basic working principles of our planet" inasmuch as they "create synergy, complementarity, supplementarity, cooperation" (Nzamujo, Personal Communication, 15/05/2020). Critical to authenticity is biomimicry defined as the attempt to model technologies on natural processes. Authenticity includes the community's bonding with its members, its technologies and its environment (Fig. 1).

In developing his principle of authenticity, Nzamujo, like both de Chardin (2004/1959) and Wilson (2012, 2014), subscribes to an ethics of moral evolution. With de Chardin, Nzamujo believes that nature itself evolves from the biosphere to the noosphere—from life to life conscious of itself. A primary characteristic of self-conscious life is empathy. Like the biologist Wilson, Nzamujo holds that *Homo sapiens* has become "*Homo empathicus*" (Nzamujo, 2018, 3) In Wilson's terms, humans like 19 other species are "eusocial" or genetically disposed in appropriate circumstances to sacrifice self- for group-interest.

An authentic community ensures the survival and flourishing of all its members. All in the community are guaranteed the means of survival: clean air and water, nourishing food, health care and basic education. A life of flourishing demands advanced education that enables community members to create their own choices about how to live in a community dedicated to the excellence of all its members. Excellence is defined by the diversity of choices that promote the future of the group's life. In an authentic community, survival and flourishing are inter-dependent and co-evolutionary.

Authenticity includes not only community bonding principles but also community connectivity to its organic and inorganic environment through sustainable technologies. Authenticity includes three spheres: ethical, technological and environmental. The ethical addresses community bonding. The technological includes the community's instruments for survival and flourishing. The environmental binds the community to its environment through sustainable technology (Tharakan 2019).

INAT members are examining ecovillage models throughout India to determine whether their ethics would incline them to become members of a global consortium of ethical ecovillages. The idea of an ecovillage has roots in the Indian independence movement with Gandhi's call to use local resources, and develop indigenous technologies (Singh *et al.*, 2019).

The history of the ecovillage community in India dates back to the establishment of self-sufficient Ashrams where residents grow their own food, engage in small and medium enterprises to produce goods and services, use renewable energy and aspire to zero-waste circular economies and ecosystems. An ecovillage community, the Muni Seva Ashram in Vadodaral Gujerat (www.greenashram.org), is dedicated to health care. The village includes a hospital, a cancer research and treatment center, and schools devoted to nursing, pre-K-12 education, vocational skills, and girls and women of all backgrounds.

The Ashram includes a guest house for patients' families and other guests, as well as residential facilities for the students in the various educational units. Grounded in the principle of sustainability, the Ashram produces renewable energy through biogas, biomass gasifiers and plasma arc recycling. Both the world's first solar crematorium and the air-conditioning systems are solar powered. While currently dependent for some power from the Vadodaral Gujerat grid, the Ashram is installing a 1.2 MW biogas plant that will render the ecovillage grid-independent and 100% renewable. Organic farming and animal husbandry provide dairy products and manure to power the biogas plants.

The proposed ethics model enjoins research and development of appropriate technologies that allow communities to develop the principles of autochthony, autonomy and authenticity to the fullest extent. In addition to ethical ecovillages, the proposed ethical model promotes the development of technologies that help remote villages move toward ecovillage model through solar installation and repair, water reservoirs, education in basic computer skills and bookkeeping and the like. Bunker Roy developed the Barefoot College in India to train grandmothers from remote villages around the world in these "ramp up" techniques. After working with Roy to develop solar villages in the Nuba mountains in Sudan, Gada Kadoda is now developing a Sudanese Barefoot College (Kadoda 2009).

An example of a community that could benefit from "ramp up" techniques is the Awra Amba Community, established in the early 1970s in the Amhara region in northern Ethiopia by the founder Zumra Nuru. The village is known as the "Ethiopian Utopia." The community focuses on the equality of its members both male and female.

With a kindergarten, schools, a library, a clinic and similar institutions, the community of 500 is a model village, combining gender equality, respect and care for children, solidarity, absence of religious rites, honesty, work and democracy. However, the village does not have access to the agroecological, energy and communication technologies of the Songhai model. The proposed ethical model would encourage the development of an appropriate technology pilot program to serve as a model for Ethiopian rural development (Eskinder 2013).

Ecovillage Versus Urban Appropriate Technology

While it's tempting, as we have done here, to focus on and work toward the possibility of green, sustainable, eco-socialist, eco-sensitive groups at the village level, such solutions are unfortunately insufficient to address, let alone solve the world's problems. Amalgamated over many groups, spanning large areas and numerous locations, they will aid in addressing planetary issues. But there are simply too many groups, too many other humans who do not live and will not be living in ecovillages.

In 2020, about 56 percent of the world's population was concentrated in the world's major cities (World Bank 2020) and by 2050, the number is expected to approach 70 percent (United Nations 2018a). In 2000, there were 371 cities with populations exceeding 1M. In 2018, the number increased to 548 and by 2030, 706 are expected to fit that criteria (United Nations 2018b). Similar to the variation among ecovillages, enormous variation in both city size, population density local environment, and ecological footprint exists (Wackernagel *et al.* 2006; NASA Earth Observatory 2013). Information elucidated from the study of ecovillages will provide keys to understanding the differences among these cities.

The proposed ethics model demands the development of appropriate technology to foster sustainability in these dense, highly populated urban areas. The ethical ecovillages hold great promise as testing sites for the recognition of best practices, for developing innovative practices, and as sites for examining scalability. Results obtained at local levels can be successfully expanded and scaled beyond the village level. Before that can be done, preliminary efforts are critical, particularly with attention to both results and design. The highly concentrated density of big cities does generally result in lower energy use and CO_2 emissions than lower population density settlements. However that their resources now come from outside the urban areas (Day and Hall, 2016).

Paradoxically, increasing urban density can diminish the threat of catastrophic climate change through appropriate technology used to develop "green" cities with reduced carbon footprint. The appropriate technologies include public mass transportation to reduce driving and increase walking (Cusick 2020; Biello 2011). Critical agricultural technologies deploy urban vertical farms in unused skyscrapers (Despommier 2010) and "green belt" farms ringing cities (Maathai 2003). Shanghai, for example, is one the world's most populated cities with 27 million in 2021 (World Atlas, 2021). Shanghai's green ring produces nearly half of its food. Compost from the city produces fertilizer and generates electricity through a landfill biogas generating plant, the largest in China (Jacobson, 2012; Decker 2010).

To reduce dependency on chemical fertilizers, Commoner has proposed a sewage pipeline system closing the cycle of food consumption in the city by returning nutrients back to the soil for agricultural production (Commoner, 1971, p.186-187). This recycling of nutrients must confront the current challenge of contamination of sewage by a multitude of chemicals including antibiotics, hormonal disrupters, etc. (Kumar et al., 2020).

The proposed ethical model envisions using appropriate technologies to promote radical changes in global physical and political economies at all spatial scales consistent with the recognition that radical changes in society are imperative to confront catastrophic climate change. Overcoming global energy poverty is critical to that challenge through rapid transition to 100% renewable energy through wind turbines, photovoltaics and Concentrated Solar Power in deserts (Schwartzman and Schwartzman, 2019). This renewable infrastructure must be created on all spatial scales from large, concentrated sites, globally managed, to small decentralized production managed by rural ecovillages.

Conclusion: An Ethical Path to A Global Community

The ethics driving appropriate technologies must be team inspired, created and driven. The teams must include all stakeholders with an emphasis on community members whose lives will be most changed by evolving values that dictate behavior change. Where possible, planning should be bottom-up rather than top-down. Historical international development often followed a top-down model. On the proposed ethics model, the interests of local community members in their own survival and flourishing must take precedence. Other stakeholders consist of 1) educational institutions; 2) government structures; 3) international agencies; 4) NGOs and non-profit organizations; 5) private sector and 6) engaged citizens outside of the local community members.

The primary step on this ethical path is the development of new appropriate technologies through research in educational institutions. The private sector is also responsible for new technologies.

Local, state and national governments must provide legal and economic pathways for the generalized implementation of appropriate technologies. International agencies like the World Bank, the International Monetary Fund, the United Nations Development Program must ensure the implementation of appropriate technologies especially in the Global South and areas of extreme poverty in the Global North.

All these agencies must work together to make sure that local communities have sufficient education to make judgments about the technologies that will direct their futures. At the same time these agencies must guarantee the continuing research, development, deployment, operation and maintenance of the technologies that their collective wisdom has declared to be *appropriate*.

References

Assmann, I. 1996. *The mind of Egypt: History and meaning in the time of the pharaohs.* A. Jenkins, trans. New York, NY: Henry Holt

Balazs & Morello-Frosch 2013. The three R's: How community based participatory research

strengthens the rigor, relevance and reach of science. *Environmental Justice* 6(1):10.1089/env.2012.0017. doi: 10.1089/env.2012.0017.

Biello, D (2011) How Green Is My City. Scientific American. https://www.scientificamerican.com/article/how-green-is-my-city/.

Cardelús, C, G Middendorf. 2013. Ecological literacy: the educational foundation necessary for informed public decision making. Front Ecol Environ 11: 330–331. http://dx.doi.org/10.1890/1540-9295-11.6.330.

Commoner, B (1971) *The Closing Circle: Nature, Man & Technology.* New York: Knopf.

Cusick, D (2020) Denser Cities Could Spare Climate but Also Increase Virus Transmission, Scientific American, E&E News on March 27, https://www.scientificamerican.com/article/how-green-is-my-city/.

Darwin, C. (1936/1859/1871). *The origin of species by means of natural selection or the preservation of favored races in the struggle for life; The descent of man and selection in relation to sex.* New York: Modern Library.

Day, JW, Hall, C (2016) The Myth of the Sustainable City. Scientific American,

https://www.scientificamerican.com/article/the-myth-of-the-sustainable-city/.

Day, J., C. Hall. 2016. *America's Most Sustainable Cities and Regions: Surviving the 21st-Century Megatrends*, Berlin: Copernicus Books (Springer Imprint).

de Chardin (2004/1959. 2004/1959. *The future of man.* Trans. N. Denny. NY: Doubleday Image.

Decker, KD (2010). Recycling animal and human dung is the key to sustainable farming. https://www.lowtechmagazine.com/2010/09/recycling-animal-and-human-dung-is-the-key-to-sustainable-farming.html.

de Luca, L., Sahy, H., Joshi, S., & Cortés, M. (2013). Songhaï, Benin, chapter 8 in *Learning from catalysts of rural transformation.* Geneva: International Labour Office, Rural Employment and Decent Work Programme. Available at: https://www.ilo.org/wcmsp5/groups/public/---ed_emp/---emp_policy/documents/publication/wcms_234891.pdf.

Desmond, A. & Moore, J. (2009). *Darwin's sacred cause.* Boston: Houghton, Mifflin, Harcourt.

Despommier, D. 2010. *The Vertical Farm: Feeding the World in the 21st Century.* New York: Thomas Dunne/St. Martins.

Eskinder, T. 2013. The Role of Gender Equality in Promoting Peace and Development: The Case of Awra Amba Community in Fogera Woreda of the Amhara National Regional State. *Anthology of Peace and Security Studies*, Vol. IV, Dec 2013 pp.238-319.

Faulkner, R. 2005/1972. *Ancient Egyptian book of the dead.* New York: Barnes and Noble.

Fortunak, J., & King, C. (2010). Building capacities for regional production of quality-assured medicines in Africa. In V. Dzidzeniyo *et al.* (Eds.), *Proceedings of the 4th international conference on appropriate technology* (pp. 173–179). http://

www.appropriatetech.net/files/Building_Capabilities_for_Regional_Production_ofQuality.pdf. Accessed 23 June 2012.

Gardiner, S. 2011. *A perfect moral storm: The ethical tragedy of climate change*. Oxford: Oxford University Press.

Haws, D. 2004. The importance of meta-ethics in engineering education. SCI ENG ETHICS 10, 204–210 (2004). https://doi.org/10.1007/s11948-004-0015-7.

Hegel, G. 1956/1837. *Philosophy of history*. New York: Dover.

Hornung, E. 2001/1999. *The secret lore of Egypt: Its impact on the West*. D. Lorton, trans. Ithaca, NY: Cornell University Press.

Hornung, E. 1999/1995. *Akhenaten and the religion of light*. D. Lorton, trans. Ithaca, NY: Cornell University Press.

Hornung, E. 1990/1982. *Conceptions of god in ancient Egypt: The one and the many*. J. Baines, trans. Ithaca, NY: Cornell University Press.

Hornung, E. 1982/1971. *The valley of the kings: Horizon of eternity*. D. Warburton, trans. New York: Timken.

Jacobson, M (2012). Shanghai urban farming. https://wwf.panda.org/wwf_news/?204455/Shanghai-urban-farming.

Jamieson, D. (2014). *Reason in a dark time: Why the struggle against climate change failed and what it means for our future*. New York: Oxford University Press.

Johnson, D. 2010. Nietzsche's anti-Darwinism. Cambridge: Cambridge University Press.

Kadoda, G. 2009. *Habboba* (grandmothers') new knowledge: Perspectives from the barefoot solar engineers project in the Nuba Mountains of Sudan. 4th WAAD (Women in Africa and the African Diaspora) International Conference, Abuja, Nigeria, August 3-8.

Klemow, K, A Berkowitz, C Cid, G Middendorf. 2019. Improving ecological education through a four-dimensional framework. Front Ecol Environ 17(2): 71. doi:10.1002/fee.2013

Kolbert, E. (2021). *Under a white sky: The nature of the future*. New York: Crown.

Kolbert, E. 2014. *The sixth extinction: An unnatural history*. NY: Henry Holt.

Kumar M, Sarma DK, Shubham S, Kumawat M, Verma V, Prakash A and Tiwari R (2020). Environmental Endocrine-Disrupting Chemical Exposure: Role in Non-Communicable Diseases. Front. Public Health 8:553850. doi: 10.3389/fpubh.2020.553850.

Locke, A. 1991. *The philosophy of Alain Locke: Harlem renaissance and beyond*. L.

Harris, ed. Philadelphia: Temple University Press

Matthai, W. 2003. *The Green Belt Movement: Sharing the Approach and the Experience*. New York: Lantern.

McEvilley, T. 2001. *The shape of ancient thought: Comparative studies in Greek and Indian philosophies*. New York, NY: Allworth.

Middendorf. G, T Mourad, J Johnston. 2020. Ecology education goes four-dimensional. Bull Ecol Soc Am, e01740. https://doi.org/10.1002/bes2.1740

Nzamujo, G. 2020. Personal communication, 15/05/2020.

Nzamujo, G. (2018). The African university of the future: A knowledge enterprise for articulating and incubating radical solutions to the formidable challenges on the continent. Ibadan Nigeria, November 15. Unpublished ms. available from author.

Nzamujo, G. 2002. *Quand l'Afrique Relève la Tête/When Africa Emerges*. Paris: Editions du Cerf.

Pinker, S. (2018). *Reason now: The case for reason, science, humanism, and progress*. New York: Viking.

Pinker, S. 2013. Science is not your enemy: An impassioned plea to neglected novelists, embattled professors and tenure-less historians. *New Republic*, August 6. http://www.newrepublic.com/article/114127/science-not-enemyHumanities. [Accessed 29 September, 2013.]

Pinker, S. 2011. *The better angels of our nature: Why violence has declined*. New York: Viking.

Richardson, J. 2004. *Nietzsche's new Darwinism*. Oxford: Oxford University Press.

Ridley, M. 1998. The origins of virtue: Human instincts and the evolution of cooperation. New York: Penguin Books.

NASA Earth Observatory. 2013. Sizing Up the Carbon Footprint of Cities. https://earthobservatory.nasa.gov/images/144807/sizing-up-the-carbon-footprint-of-cities.

Nietzsche, F. 2011/1881. *Dawn*. Trans. B. Smith. Stanford: Stanford University Press.

Schell, J. 2000. *The fate of the earth* and *Abolition*. Palo Alto, CA: Stanford University Press.

Schwartzman P, Schwartzman D (2019) *The Earth is Not for Sale: A Path Out of Fossil Capitalism to the Other World That is Still Possible*. Singapore: World Scientific.

Sen, A. 2000. *Development as Freedom*. New York: Anchor.

Silversmith H, Ruggles D, eds. *Cultural heritage and human rights*. New York: Springer; 2004.

Singer, P. 2011. *The Expanding Circle: Ethics, Evolution and Moral Progress*. Princeton, NJ: Princeton University Press.

Singer P, Stavenhagen R. 2004. Cultural rights and human rights: A social science perspective. In: Pitarch P, Speed S, Solano X, eds. *Human rights in the Maya region: Global politics, cultural contentions, and moral engagements*. Durham, NC: Duke University Press; 2004.

Singh, B., Keitsch, M., & Shrestha, M. (2019). Scaling up sustainability: Concepts and practices of the ecovillage approach. *Sustainable Developmen*t, 27(2), 237–244.

Tharakan, J. 2019. The Songhaï Agroecological Sustainable Development Model: Synergy, Symbiosis, Collaboration, and Complementarity. In *Regional Development in Africa*, ed. Norbert Edomah. Rijeka, Croatia: IntechOpen. http://dx.doi.org/105772/ intechopen.87929. https://www.intechopen.com/online-first/the-songhai-agroecological-sustainable-development-model-synergy-symbiosis-collaboration-and-complem.

Tharakan, J. 2017. "Indigenous Knowledge Systems for Appropriate Technology Development." In Indigenous People, edited by P. Venkatesan, 123–134. Rijeka, Croatia: IntechOpen. doi:10.5772/65629.

Tharakan, J., Castro, M., Trimble, J., Schwartzman, D., Stephenson, B., Broome, T. & Verharen, C. 2006. Diversifying engineering education: A seminar course on the ethics and the philosophy of appropriate technology. *Global Journal of Engineering Education*, 9:2, 111-119.

Tharakan, J., Mitchell, M., Moore, J. & Jenkins, G. 2008. Design of a combined solar energy system for a remote flux tower and a rural community. In J. Tharakan, J. Trimble, eds., *Proceedings of the. 3rd international conference on appropriate technology*, 51-58. Kigali, Rwanda: Ministry of Science and Technology.

United Nations. 2000. United Nations Millennium Development Goals. https://www.un.org/millenniumgoals/

United Nations. 1948. Universal declaration of human rights. http://www.un.org/en/documents/udhr/index.shtml. [Accessed 23 February 2013].

United Nations. 2018a. The World's Cities in 2018 Data Booklet. https://www.un.org/development/desa/pd/content/worlds-cities-2018-data-booklet.

United Nations. 2018b. Population Prospects. https://www.un.org/development/desa/en/news/population/2018-revision-of-world-urbanization-prospects.html; see also https://population.un.org/wup/).

Verharen, C., Bugarin, F., Tharakan, J., Wensing, E., Gutema, B., Fortunak, J., Middendorf, G. (2021). African environmental ethics: Keys to sustainable development through agroecological villages. *Journal of Agricultural and Environmental Ethics* 34(18), 1-18.

Verharen, C., Gutema, B., Tharakan, J., Bugarin, F., Fortunak, J., Kadoda, G., Liu, M., & Middendorf, G. (2014a). African philosophy: a key to African innovation and development. *African Journal of Science, Technology Innovation and Development*, 6(1), 3–12. https://doi.org/10.1080/20421338.2014.902565.

Verharen, C, Tharakan, J, Bugarin, F., Fortunak, J., Kadoda, G. & Middendorf, G. 2014b. Survival ethics in the real world: Research universities and sustainable development. *Science and Engineering Ethics*, 19(2), 599–623. DOI 10.1007/s11948-013-9441-8.

Vodouhe, S. & Zoundji, G. (2013). The Songhaï Centre in Benin. *MEAS Case Study Series on Human Resource Development in Agricultural Extension*, #7, January. http://www.measextension.org/meas-offers/case-studies. Accessed 24 June 20.

Wackernagel, M, J Kitzes, D Moran, S Goldfinger, M Thomas. 2006. The ecological footprint of cities and regions: comparing resource availability with resource demand. Environment & Urbanization 18: 103-112.

Wilson, E. O. (2014). *The meaning of human existence*. New York: Liveright.

Wilson, E. O. (2012). *The social conquest of nature*. New York: Liveright.

World Atlas 2021. The Largest Cities In The World By Population. https://www.worldatlas.com/articles/the-10-largest-cities-in-the-world.html.

World Bank 2020. Urban Populations. https://data.worldbank.org/indicator/SP.URB.TOTL.IN.ZS?end=2020&start=2018

Chapter 10

AT and Innovation: A Perspective for African National Structural Transformation based on a Networked African Innovation System

Mammo Muchie and John Trimble

Abstract

How to create the inventive, transformative, innovative, green, smart African integrated, networked and unified industrial structural sustainable development path to deal with and respond to the fourth and fifth industrial revolution is the challenge of the time for Africa. The world is now at the stage moving from industrialization to digitalization. How can Africa build an integrated, networked, and unified innovation system for sustainable development to undertake catching up by navigating the complexities of the current global value chain in this digitalization time? The current digital technology provides both opportunities and challenges. Can Africa leapfrog into the digitally driven knowledge economy to remove any risks and dangers to achieve the fully developed status without losing as Africa did in the first, second and third industrial revolutions? There should be research on how to apply transformative innovation and sustainable development to the African context. No imitation of national innovation for Africa. No imitation of development economics for Africa. Africa needs a contextually anchored new model that synthesizes innovation with an African integrated and sustainable development innovation system. This system must embrace technologies that empower people. The chapter will explore how to create an all-inclusive system of innovation to promote fully the African sustainable development by applying African agency, identity, project, freedom and unity.

Introduction

A brief overview of the African economic picture reveals a paradox where the continent that has rich mineral resources, over 1.4 billion people and a land mass which exceeds the sizes of China, USA, India, Western Europe, Argentina (see map fig.1) together. Africa while larger than the sum of these regions is in the an unacceptable state of being an object of aid, debt and loans. The African continent holds vast resources both known and yet to be explored in the post- colonial period. Africa should have been a production and innovation center not a charity and aid center of the world where still since 1963 when over 30 states achieved formal independence, 'donor ship' has replaced African national ownership' of not just Africa's resources, but even more worse Africa's own agency, autonomy and independence to shape policy and direction to undertake the integrated sustainable African innovation system. The main thrust of the African quest to unite both the politics and the economy flows from a recognition that Africa must organise a production, economic and innovation system by integrating producers with producers, producers with users, users with users in Africa and for Africa. The development and deployment of technology and techniques that empower local communities is required to achieve this. It is both desirable and possible. Knowledge of how to do it- the know- how innovative knowledge imagination- must be assiduously cultivated. Only then will the continent emerges fully as a region free from the donorship gaze it suffers from.

It is no exaggeration to state that today's African political and economic arrangements are characterized by internal pervasive and schizophrenic disconnections, mismatches, fragmentations and external dependence. Nearly 70 % of Africa's overall population exist in subsistence and primary resource and agrarian condition. Youth unemployment continue to grow rather than decrease. Where a country has the overwhelming portion of its production as agricultural, that country invariably remains vulnerable even to feeding itself, conversely where a small percentage of a nation's population is working on the land and is engaged in manufacture and services for the most part, that country is more likely to feed itself whether it rains or not.

The main pattern of Africa's economic relations with the world economy is through what makes Africa permanently remain vulnerable unless it changes its conditions through continued unequal primary agriculture and mineral exports for the products manufactured elsewhere. African countries produce similar primary products for the same market and compete against each other thus accentuating and deepening their fragmentation making inter-African trade the lowest in the world.

Africa faces a true dilemma. If it is able to insulate itself from the world economy fragmented as the states are at present, it can incur income and knowledge losses. If it integrates on current dominant patterns of relating based on primary commodity transactions with the world economy, it faces continued economic dependence and fragmentation and lack of structural transformation of its fundamental economic structure.

Africa's current pattern of insertion in the world economy comes at the cost of fragmenting the African economic and political space. It appears the continued cost of fragmentation is supposed to be offset by Africa being in the international aid system. Whether African fragmentation can be offset by dependence on aid or national development should be a genuine issue for investigation and reflection.

The problem is that after nearly 50 years of post- colonial independence African economies continue to be fragmented in spite of the AU/NEPAD salutary processes. The more the fragmentation amongst African economies deepens the harder for each of the fragments not to be supplicants to the aid system. Africa thus also faces another critical dilemma of being addicted as a long-term victim of the burgeoning aid industry that has created a business of what is known as 'technical assistance' where those who provide the aid consume mostly the resources allocated, and the recipient Africans continue to be in a vulnerable position as long-term aid receivers without a way to get out of this dependency situation. "Charity corrupts, long-term charity corrupts long-term." [1, p. 12] In general, it may not be easy to disprove aid is not useful to some within the recipient countries. However, this does not validate aid or the aid system, since it is not also difficult to show, that the long term impact of aid is negative, if we proceed from the normative preference that the recipient countries options to plan their development free from conditions imposed that often do not take the specific context of the countries can be misdirected by aid system. Tanzanian leader that passed away recently announced to all the donors that they want time to think and cope with the influx of hundreds of donor inputs. They said they needed time to work out what this all means and made a moratorium on donors' visits to Tanzania!

It is thus no exaggeration that a country relying on aid is most likely not to develop a national strategy without the interference and the factorings of the interests and policies of the aid system. Being a recipient in an international system for many African countries has not brought development but instead corruption and poverty. It undermines the country from making mistakes and learning from the routines and practices of the possible economic evolutionary activities. Africa cannot afford to continue to suffer the opportunity cost from continuing to receive aid only to defer building the much-needed ability to create the capacity, capability, competence, learning and innovation to transform the largely agrarian and subsistence economic system with the current digitalization knowledge economy anchored innovation system by skipping the stage of industrialization.

Much of existing development theory is of no help when it comes to addressing with a root cutting perspective of how Africa might increase its possibilities given impossible constraints on its transformative potential the existing pattern of relationship- fragmentation needing the aid system, and conversely the aid system deepening African economic fragmentation imposes.

A critique of the development theory that has evolved over the past half century takes as given this myopia of the 'fragmentation-dependence'(F-D) dilemma or situation that account for most of what is proverbially known as the unchanging African predicament, or crisis. There is a need for a fresh approach, a new departure to embark on an intellectual search for concepts, theories, and strategies to convert the 'fragmentation-dependence' dilemma into an enabling 'integration-self-sustaining capabilities building' system to re- launch African development on a secure pedigree.

It is urgued that the challenge of addressing fragmentation and dependence whilst learning to integrate the African economy on a new basis with the world economy requires the rethinking of mainstream development theory as it currently stands. One way in which this re- orientation can be advanced is through the appropriation (not mimicry) of the knowledge that has emerged from evolutionary economic theory and the application of the national system of innovation framework for an integrated African system of economy, production, and innovation. And sustainable development.

The chapter will try to undertake a brief critique of existing theories that have informed policy in Africa in order to define a new research agenda to provide metaphors, heuristics, alternatives and the resources for an emancipatory epistemology and practice in the rebuilding of African integrated sustainable development Innovation system by unifying all the social, economic, education, health, environment achievements with full success and no failure.

Overview of the Problem for Theorizing African Development

We ask the question of how development economics has framed or miss-framed the development question in Africa. Has it been a help or a hindrance? Why is that after nearly half a century the continent has no substantial model of success to show as we see in East Asia, and now China, India, and in Latin America such as Brazil? Was the knowledge needed for changing and structurally transforming

African society being developed or was there something flawed in both the production and utilization of this knowledge that technical assistance motivated development economists injected in Africa without any significant results to show for it in the end?

Figure 1: Map of Africa larger than China, Europe, India and USA combined

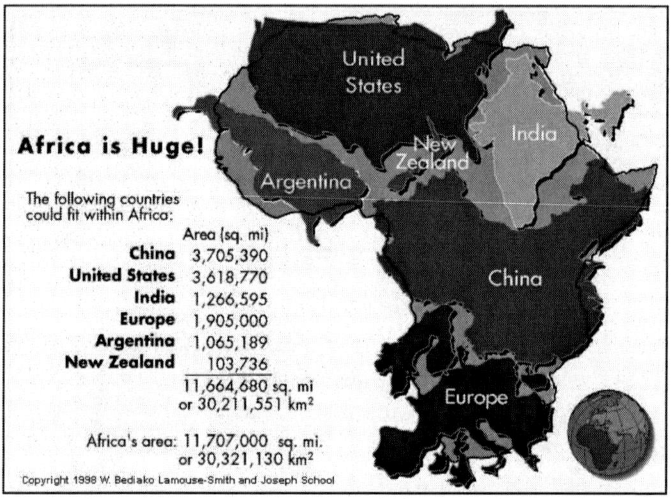

Has development economics and development theory in general addressed the problems of structural transformation of developing economies by building in the overall dynamics of change knowledge, innovation, learning, and competence building in the process of economic development by studying the processes of economic change, and the economic and non-economic institutions and incentives supporting or hindering the transformation of varieties of social economies. It appears that much of conventional development economics lacks both the 'words and the grammar' to use a Heideggerian metaphor to address the being and process of social-economic change. The conceptual preference to use evolutionary economic theory together with studies of economic and non-economic institutions, markets, policies, incentives, knowledge, innovation and learning, competence and capability creations, accumulations, diffusions, flows and implementation processes is in part to find 'the words and the grammar' in order to develop fresh insights on the problems and challenges of economic development in Africa. The goal should be to move beyond the tinkering of existing arrangements characterized by deep fragmentations and dependence to overcoming these and find the know-what, the know-why, the know-how to understand the technologies, routines, and practices for the making of an integrated and independent African sustainable innovation system. A critique of development economics is followed by the suggestion that both evolutionary economic theory and the system of innovation framework may be appropriated usefully to re- imagine and re-articulate an African system of economy, production, and innovation.

Briefly, according to Richard Nelson, evolutionary economic theory sees the economy in the process of change and holds that economic activity cannot be assumed to take place in an economic context already familiar and also readily understood to economic agents. This contrasts with neo- classical theory that views the economy in a state of rest and assumes a context where agents make decisions based on relevant experience or on past knowledge. Formal modelling for evolutionary economic theory begins from assumptions of dynamic systems that may be far from the state of rest, while neo-classical theoretical models assume that a state of rest will or can obtain always. Both also differ in the way they understand economic behaviour with evolutionary economic theory introducing to the rationality of agents bounded and creative and innovative behaviour, while neo-classical theory assumes agents face given and familiar choice sets allowing them the opportunity to make optimum choices.

The application of an evolutionary economic theory to generate insights for the making of an integrated transformation of largely an economically fragmented Africa into a modern economic system invariably throws up challenges. The process will be complex throwing up both myopia and varieties that require evolutionary mechanisms of selection and self-replication. The process of integrationisItselfevolutionary.Evolutionaryeconomicstheoryhas built a body of knowledge on evolutionary processes. The creation of distilled knowledge using evolutionary economic theory for integrating the African economy by opening various options will help to probe the trajectories of Africa's overall transition from the ongoing fragmented structure to enter into the African unified knowledge economy digitalization time.

Evolutionary economic theory will be used to capture the key parameters that can articulate as real a picture as possible regarding what is taking place in the African fragmented and disarticulated real economy in order to identify the key variables and determinants that can stimulate an overall better economic performance and economic progress. African economies are fragmented, are largely skewed, and lack inter-African integration. Both economic and non- economic institutions are weak in Africa. The interaction of markets and policies, or economics and politics, the internal and the external, the domestic and the international has a lot to be desired even in the better economy performing parts in Africa. There is thus a need to undertake a radical reconceptualization and attempt to create a fresh body of high-quality scholarship on African economic growth by considering the co-evolution of technologies, firm and industry structures and supporting economic and non-economic governing institutions in each African economy starting with the most advanced economy of South Africa.

The innovation system framework is an offshoot of evolutionary economic theory applied to the organization and interactive processes of the significant variables that generate economic development. The framework suggests that theoretically development is not only a process of production but also it is a process of knowledge, innovation, learning and capability/competence building. This means developing economies cannot grow only by using already produced knowledge, but they must also be involved as producers of knowledge. In Africa, the challenge is to create economies that are driven by indigenized systems of innovation. The question of how to bring this about and opening basic research programme to identify the significant determinants and examination and mapping of the co evolution of technologies and institutions become worthwhile. In the African context, there is a need to begin with a skepticism that viable and functioning integrated sustainable development innovation systems are still to be constructed. Even in the economies such as South Africa, they exist in bifurcated, lop sided and fragmented patterns. Intuitively, in many other African economies they are either dysfunctional or weak, or hardly no-existent. This makes the challenge a different one: rather than studying an already existing and well-developed system of innovation, institutions and technologies, the intellectual challenge becomes one of how they can be created and sustained given the African environment as a constraint.

The approach taken here is exploratory and not conclusive choices of theories. The idea is to search and look for fresh insights to address how an evolutionary economic theory and systems of innovation framework perspectives together can shed new light in making an integrated African innovation system of production. What are the likely evolutionary paths, variations and selective mechanisms that can facilitate manifold inter African interactions, producer to user interactions, user to user, user to producer, and producer to consumer, and producer, and producer to knowledge creators and public policy makers that spill over borders that are, extend over the existing arbitrary boundary constraints in Africa.

From fragmentation to integration, from dependence to independence and from lack of innovation to innovation, requires searching for appropriable models as metaphors to create knowledge that can be fed into policy to yield tangible impacts in helping change the reality on the ground. After fifty years of disintegration, conflict, and poverty, it is long overdue that Africans begin to think out of the box in order to stimulate new thinking and policies whose impact will make a difference.

Development Economics and Its Problems in Relation to African Development

It has been claimed that just as economics can be defined by what economist's do, so also can development economics be defined by "what development economists do" [2, p. 8]. The interesting issue is what did development economists do? Some development economists accept that 'development economics has not produced the results expected of it" [2, p. 9]. According to Hirschman the problem, in part, has to do with the expectation of performance from developing countries: the latter were 'expected to perform like wind-up toys and 'lumber through' the various stages of development single- mindedly... these countries were perceived to have only interests and not passions [3]." Arthur Lewis attributed it to the lack of historical perspective of economists [4]. Krugman attributes the problem to lack of rigour and claims that what remains heuristically valid from the development economic theory and modelling is the following: external economies, increasing returns, complementarily between sectors and linkages [5]. Reflexivity by the economists about the work they do should be commended. But what all of them did not address is that development economics framed development of the agrarian transition under a capitalist political economic framework. It arose in part in response to what seemed to be a relative success by those who framed development not under the capitalist from but in opposition to it. The charge by advocates of mono-economics, why there is a need for a separate economics is in part related to the fact development economics ought n o t to be radically different from any other economics since much of the problems and themes relate to developing within capitalism and not in opposition to it. Development economics set itself for attack by both those who argued that the transition and transformation of developing countries may be hundred by capitalism and those who see no hindering from capitalism. Development economics could have done to capture a domain that can justify its special identity by framing the challenges of development and transformation as national whether that national is framed within capitalist or other variants of political economic system. In the African context, if there is a criticism of development economics it is because it is not Listian-Schumpeterian and more over the national in Africa needed to be understood beyond the existing states in Africa.

Hirschman has a point that it would not be complete to take the assumed interests of developing countries and expect economic development to be the outcome. It is also critical to understand and contextualize interest with national passion. Development economics ignored national passion. In Africa the key matter is how to overcome the dependency and colonially sanctioned and legitimized and post- colonially bequeathed political, economic fragmentation into account along with the passion to remove these by bringing about comprehensive integration and capability to deal with the technical assistance or donor- ship dependency imposition and continuation for unlimited time. Lewis is right to raise the lack of historical perspective, but this is not simply a matter of lack of history learning by economists, it is more the lack of historical contextualization to address the real challenges to achieve Africa's full development. The lack of contextualizing development economic theory to the specific conditions requiring change and transformation is what made much of the insights from development economics not useable both to policy making and real impact with tangible results for Africa.

Krugman's point of lack of rigour does not address the core dilemma: what difference does pushing externality costs matter to those who produce the same commodity for export by undermining each other's producers' world market prices on the primary goods that fetch them the countries principal foreign exchange? What the quotes from these three eminent development economists show is development economics has problems. What each has observed to be the problem does not, however, go far enough to address the core issues that block transformation and poverty eradication in Africa.

The rationale for moving from economics to development economics is to highlight the fact that development challenges perforce involve the marriage between theory and the reality in the analysis of the condition of underdevelopment to the transition to development.

There were three sources for developing theories: the first is to draw upon the classical economists such Adam Smith, David Ricardo, Thomas Malthus, John Stuart Mill, William Petty and

Karl Marx whose main concern was to understand the causes and determinants of wealth making and economic progress. The second is to draw upon the experience of the currently industrialized countries that have undergone structural transformation from an agricultural economy to an industrial economy by making Africa lose the first, second and third industrial revolutions. The third is to address the context and issues and not overlook them for the underdeveloped regions themselves. Where it seems development economics did not succeed at least in the African case is to fail to historicize and contextualize Africa's development dilemmas and assume that reworking and adapting classical theorems and lessons from the developed economic advance will provide the tools, models, metaphors and heuristics to inspect the specific development challenges of developing economies suffering from dependency and fragmentation at least in the African case.

The various models that were generated from development economic theory and research are welcome contributions to see different aspects of the development and economic growth problematic: low level equilibrium trap, the big push, dynamic externalities, dualism, circular and cumulative causation, growth pole, rural-urban migration, impact of population on economic growth and the vice versa, social cost benefit analyses, immiserating growth, dual gap analysis, missing markets and rent seeking and so on [6].

The core themes of development economics have changed over time. In the 50s and 60s which Paul Krugman calls the years of "high development theory [5]" and with examples of successful extensive industrialization in the former USSR the role of the state, import substitution industrialization, planning were emphasized amongst the choices of development themes. Analysts put the influence of the former Soviet Union's efforts at industrial development as follows:

> "It should never be forgotten, however, that no country in the world ever made such an economic advance in such a short space of time as the Soviet Union did after 1918, through a planned allocation of resources that favored

> investment at the expense of consumption. The fact that planning may be operated too rigidly, or for too long and go wrong, should not be allowed to obscure the fact that it also has merits, and that unfettered free enterprise can also lead to economic disaster and social deprivation [2, p. 6]."

Some of the themes that early development economists addressed include the following: rural underemployment and disguised unemployment, late industrialization, the utilization of underemployed manpower, acceleration of capital accumulation, deliberate intensification and guided effort, and new rationality for protection, planning and industrialization.

The development strategies related to the development themes above were the following:

a. industrialization
b. Rapid capital accumulation
c. Mobilization of underemployed manpower
d. Planning and an economically active state

The early development economy theory equated development with growth and industrialization as developing countries were seen as 'primitive versions' of developed countries with time to grow into the civilized status of Europe! The rapid industrialization of the former USSR was an example serving a lesson not to ignore state intervention even though not as totalitarian as it was in that part of the world.

There were criticisms not to treat social consequences simply as appendages of underlying growth theories. Industrialization was seen as unsustainable if it comes with human cost (e.g. population, inequality, urban bias, unemployment). There was a move not to treat education, health and other issues subordinate to economic growth. Later populist theories like that of Schumacher questioned economic growth theories altogether. The notion that the development of agrarian economies can be considered as backward version of western economic development was also challenged by structuralists who suggested structural problems are both distinctive and specific requiring differentiated modelling and explanatory structures. What is interesting about development economics is that the normative stance of the particular theorist clearly informs the choices of issues, themes and remedies preferred.

In the 80s arguments from the Right stressed the neglect of the market and the unacceptable state activism in the economy as the problem for economic development, not economics, not the market not equilibrium or neo-classical economics. In fact the neo-liberals argued that more of neo- classical economics, less development economics is the way forward for dealing with growth and poverty problems. In fact, Hirschman explained the "alleged demise of development economics" to the "resurgence of neo-classical orthodoxy" on the one hand, and the "neo-Marxists who reject the claim of mutual benefits "from the integration of developing countries in the world economy [2, p. 7].

Amartya Sen came up with an argument of taking the issue beyond the simple state and market dichotomy and opening development to themes to what may prolong the life of development economics. His was work that extended the thematic choices for development economics by adding development as freedom, development as entitlement, development as democracy and finally development as capability. But the theoretical challenge is to provide a general framework for an understanding of structural transformation and development, and at the same time provide an adequate explanatory power on the nature and dynamics of the transformation processes.

What Amartya Sen did was however concentrate on extending the choice set for themes and issues that economic development can incorporate. Sen admitted that traditional development economics may not have been as dismissible as various trends suggest. Instead, he acknowledged what he calls 'thematic deficiency' at the core of development economic theorizing. He argued concentration on national product, aggregate income; total supply of public goods, capital accumulation, and the creation of surplus may be necessary but not sufficient to get development that includes the least advantaged. He stressed entitlements of people and capabilities these entitlements generate are the relevant thematic choice for development economics.

Sen said the process of economic development has to be concerned with what people can and cannot do or to use his words have reason to value and choose to do! Development economics has bifurcated into neo-liberal versions (Poverty of Development Economics, 1981) and its critiques (The Dilemma of Development by John Toye, 1987) on the one hand, and the Amartya Sen. thematic extension from national product to entitlement, freedom, democracy, and capability leading to human development and inclusions of social specific themes such as gender, deprivation, hunger, basic needs and environment.

What is missing with both filling in the missing themes or extending the list is the very framework for conceptualizing the economic progress of nations and how their national transformation can be theorized with a view to understand and explain the pace and rhythm of the development process during the transition from largely underdeveloped socio-economic historical existence to a post-colonial fully developed economy and liberation. The fundamental categories of analyses can be illuminative and insight generative when the analyses start from the specific historical experience with a normative commitment to generate knowledge for policy with relevant impact to bring about development, transition from underdevelopment to full development.

Can Evolutionary Economics Theory succeed where development economics theory fail?

"Reconsidering the history of development economics provides valuable opportunity to challenge (orthodoxy) [7, p. vii]." Orthodoxy has spawned what has been described as the "New Development Economics [7, p. viii]" or what others call the neo-liberal development counter-revolution. The debate has gravitated between the Washington and Post Washington consensus or showing the history of economic ideas is the history of economic development and not the abstract formalization of neo- classical static equilibrium economics. There is a third option where evolutionary economics can either align its framework with development economics and refine and appropriate its analytical categories to address beyond issues and themes of development how a framework for the pace, rhythm and dynamic evolution of development takes place, or development economics can renew itself by incorporating or redefining the development question with evolutionary dynamics. Both possibilities are open and should be pursued. Development economists should reach out with a view to revitalize and redefine the whole framework of development economics on a new axiological foundation. Conversely evolutionary economists should reach out to development economics and find ways of incorporating some of their insights to redefine the development framework in to a new and more robust intellectual foundation.

Evolutionary economics built its metaphors from evolutionary biology as new-classical economics tried to build its metaphors from physics. What is needed is an economic theory or a synergy of economic theories that can build their models and metaphors from history, specific context of countries in transition from the underdevelopment condition to a fully developed condition, from fragmented states into the transition to an integrated state, from a state of dependency to the transition to independent states to discover insights for a combined solution to the problem of heterogeneity and change.

Evolutionary economists built evolutionary economics mainly as a reaction to neo- classical economics. As Krugman puts it: They reacted as follows:

1. They want to get away from the idea that individuals maximize. Instead, they want to represent decisions as the result of some process of groping through alternatives, a process in which it may take a long time to get to a maximum - and in

which the maximum you find may well be local rather than global.
2. They want to get away from the notion of equilibrium. In particular, they want to have an approach in which things are always in disequilibrium, in which the economy is always evolving. Lately there have also been some economists who want to merge evolutionary ideas with the Schumpeterian notion that the economy proceeds via waves of "creative destruction".
3. Their field is about the interaction of self-interested individuals - who are often thought of as organisms "trying" to leave as many offspring as possible, but which are in some circumstances best thought of as genes "trying" to propagate as many copies of themselves as possible. The main difference between evolutionary theory and economics is that while economists routinely suppose that the agents in their models are very smart about finding the best strategy - and an economist is always defensive about any model in which agents are assumed to act with less than perfect rationality - evolutionists have no qualms about assuming myopic behaviour. Indeed, myopia is of the essence of their view [8].15

Nowhere do we find agents with myopia than in Africa that includes both externally, i.e., those who sell charity to Africa whilst allowing continuing plunder of its natural resources and the elites that are often bereft of an African national spirit and consciousness unable mostly to develop a national development framework for Africa's transformation.

Some of the features that can be incorporated also to a development economic framework from evolutionary economics are as follows:

1. Stimulate and understand inter economic and non-economic actor interactions and dynamics
2. Co-evolution of economic and non-economic governing institutions, practices and understanding (Richard Nelson)
3. The interaction of policies, knowledge, incentives, institutions, practices, and the understanding involved in the process

4. System building, to identify significant interactions and interfacing of parts
5. Bridge the gap between theory and reality
6. The sources and organization for stimulating innovation, imagination and creativity, how routines are formed, and novelties emerge and selections are made.

The review of development economics theories and some of the thinking of the leading theorists shows that the project to renew development economic theory is satisfied by adding new themes and selecting others. This, however, is inadequate when it comes to African's transition from an agrarian condition to an industrial economic condition. The failure of development economics in Africa has nothing to do with choice of themes; it has everything to do with failure of framework. It is not thematic deficiency as Sen puts it, when it comes to the peculiarities of the African condition, it is rather framework deficiency, i.e., the central framework for understanding and explaining the dynamics, pace, rhythms, and contradictions of the African development processes.

Why evolutionary economics may be useful in the African context is, if it can provide a framework for understanding the evolutionary dynamics of the African economy, which influence and impact states of social-economic system and given its main concerns relate to the transformation of generic ideas, social or technical knowledge with the principal overarching theme of integration rather than tinkering with the existing fragmentation. We suggest that integrating Africa or making the Africa nation itself is an evolutionary problem of dynamics, of creative destruction, requiring evolutionary approaches to understanding and creating knowledge. Evolutionary economic theory has such concepts that can be useful to stimulate research in making an integrated African economy. The African economy is in a state where it exists with a non-linear, chaotic, complex economic phenomena where the need to make products, technologies, innovations, institutional arrangements and incentives is anything but compelling. In dealing with the problem of transition and transformation of the African economy we are dealing with evolutionary processes. In Africa change is critical dependent on initial conditions and path dependence. Now it may be mainly destruction, Africa needs to enter the Schumpeterian process of creative destruction to make the African the innovation system with creative reconstruction [9].

National Innovation Framework

National innovation system (NIS), we would argue, is not just a tool to achieve the narrow goal of industrial/economic competitiveness, but it is about achieving a broader development and wider social benefits for a people and nation. Major elements of NIS can be identified in figure 1 below as:

1. Conceptual Framing: that is, ideas, policies need to be linked to a conceptual framing of how economics and politics are co-governed and/or co-evolved. And responding to opportunities and dealing with challenges require policies to be rooted in a conceptual framework that governs the dynamic interaction between a nation's political and economic change.
2. Institutions, Technologies, and Knowledge: This needs strong interaction, linkages, synergies, and co-ordination to achieve coherent co-evolution leading to an efficient innovation system and higher level of technology/knowledge accumulation.
3. Incentives: Appropriate incentives to achieve co-evolutionary dynamics between institutions, technologies, and knowledge production by linking economic and non-economic agents to meet stated goals and objectives.
4. Implementation/ Learning/ Feedback/Outcomes: Implementation of strategies, policies, projects, and programmes should include feedback mechanisms (review, monitoring, and feedback) leading to learning outcomes. Ability to learn - self learning and ability to take corrective measures are imperative for building technological capabilities and embed innovation dynamics in both narrow industrial and broader socio-economic development. This is where implementation/practice feeds to learning outcomes and, conversely such learning feedbacks in turn feedbacks to generate deeper understanding to stimulate further and hitherto unexplored innovative

> practices. Socio-economic Changes: Learning outcomes could lead to different types of socio- economic changes – corrective, adaptive, evolutionary, structural, contingent, and so on. Progressive Transformation: When structural change results in the most dramatic positive change of political/socio-economic system (but it is unlikely to happen often); Regressive changes: When transformation results in negative rather than positive impact on political/ socio-economic system leading very often to conflict and social-economic tensions and breakdowns.

It matters, therefore, how structures and agents interact and co-evolve in dynamic systems for the type, scale and quality of transformation that can be undertaken in specific contexts and situations. It is when all the above identified structures and elements (both economic and non-economic) enter into specific relationships in given spatial, economic, political, social, historical and cultural contexts to facilitate change or breakdown that provide the rationale to find a conceptual tool that can capture the salience and core dynamics, rhythms and paces of development processes. Innovation conceptualized at a nation-state level has been described as a national system of innovation. This concept has been described as the national political economy of production (List, 1852). The merit of this conception is to bring together four essential elements: space, economy, politics, and knowledge and how their interactions, synergies and systemic combination generate transformation. The way the elements interact in given historical contexts that nations are in can either lead to dynamic efficiencies or inefficiencies. It matters therefore how the elements are linked and co- evolved to achieve an efficient innovation system leading to higher level of technology accumulation and economic development. This is captured by Figure 1.

The efficiency of the NIS is determined by effectiveness of key factors displaced in Figure 3 below. They are infrastructure, relations and linkage, investment and knowledge and talent. As figure 3 indicates all four of these factors impact each other. The nature of these factors and the impact they have on each other determine whether the NIS will be efficient or inefficient.

Figure 1: Major Elements of National Innovation System (NIS)

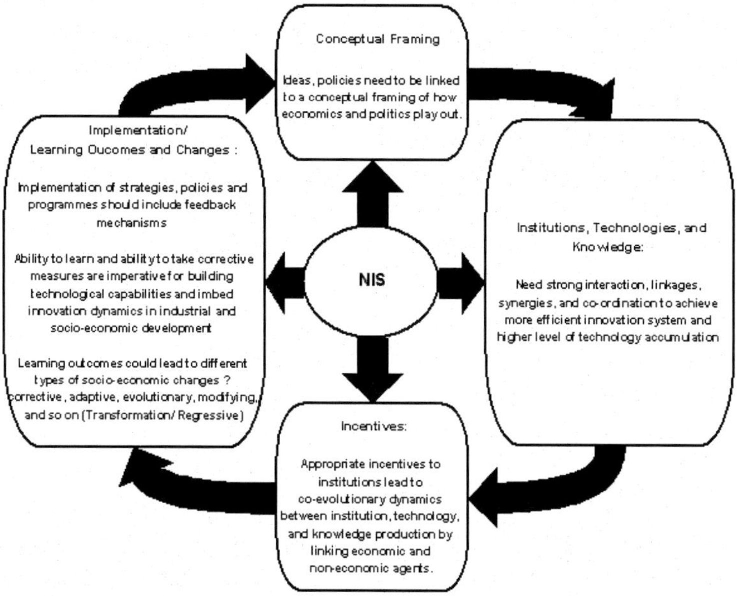

Figure 2: Linkages between Institutions, Technologies, Knowledges, and Incentives in NIS

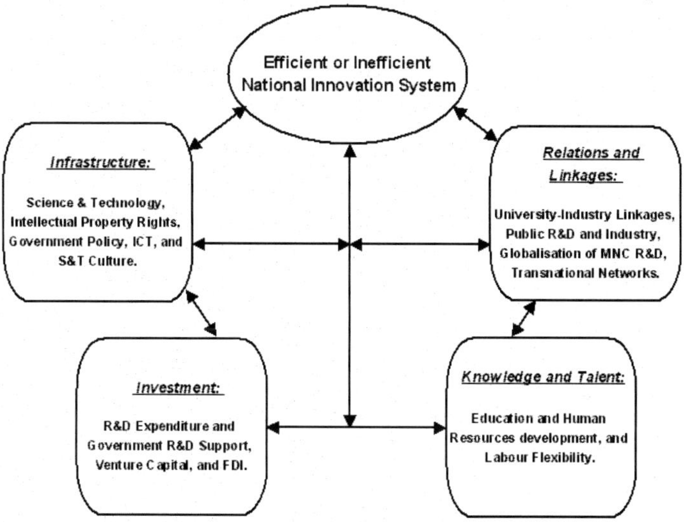

From a developmental perspective there is no doubt a need to establish a unified conception of the system of innovation where the politics and economics, the internal and the external, the domestic and the international interactions are duly included. A political economy perspective of systems of innovation is critical to develop to shed new light and insights on the development, transition, and transformation processes of agrarian societies into manufacturing nations.

A comprehensive model is presented in Figure 3. It unifies the context of both the political and economic domains. Both domains are interlinked through knowledge, technology and research. Innovations are expressed as products, processes, organizational and institutional advances. Innovations take place along four dimensions: space, technology, sector, and firm (organization scope). Space refers to local, national, regional or international geo-space. Nano-technology, bio-technology and ICT are examples of the technology dimension of the innovation. The four major sectors include agriculture, services, manufacturing, and mining (along with other natural resource retraction such as oil and timber). The major types of organization scope are multi-national corporations, small to medium enterprises and State-owned enterprises (SOEs). This model that unifies the key variables and their interactions is critical to identify and arrange in order to understand and explain the factors and dynamics of the development process with fresh insights.

Integrating the African Economy: Bridging Rhetoric with Practice

On May 25 in 2023, it will be 60 years since the Organization of African Unity was formed, yet the African economy is neither organized nor united. This is despite the various attempts that have been made by the officials in Africa to unite the continent's infrastructure and economy. As expected, they agreed to form a commission how to build African unity rather than going for unity itself.

Figure 3: Unified Model of System of Innovation

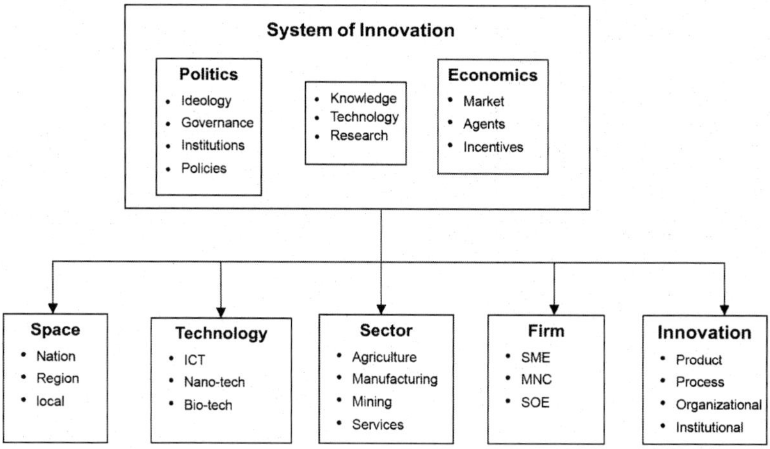

We must be aware that all times are times of transition, but after 500 years of oppression, over 100 years of pan-African revival, 50 years after Ghana's independence, nearly 40 years of the OAU, and 11 years of the AU, the current moment is not a moment to say let us be cautious, let us be incremental, let us go slow and let us not speed up and get on the slow lane of history. No, the moment requires that Africa should not throw its destiny as hostage once again to the misfortune of the slow pace and the incremental. Time and time again we have seen such calls for the slow lane have become the wrong lane of history. The peculiarities and specific condition in Africa call exactly for the opposite to the slow that has also become unfortunately and invariably very often the wrong lane. If Africans go slow and one by one as we are certain a number of countries may wish or are likely to prefer, they will always end up being open to one form of interference to another, one form of conflict to another.

It is very compelling and urgent to have all the commitment to realise a Pan-African unity to bring about policy and strategy coordination to realize unity. There is a need to use an African integrated sustainable development national system of innovation perspective to deal with the theme of the making and organizing of the political economic unity of Africa. Without any further delay [10].

Table 1: Summary of Stakeholders role in promoting innovation

Stakeholder Category	Recommended actions
Education	K-12 education should address innovation in the broad context of critical thinking with hands-on engagement linked to community needs, university research and curriculum should address innovation at all levels across all disciplines linked to collective engagement for social advancement instead of individual profit
Government	Intellectual property laws must be shifted to benefit communities instead of individual and corporate profits. Funding for collective community-controlled innovation must be promoted
International Agencies	Regulate trade and technology transfer to address inequality. Focus funding on innovations that assure basic needs and infrastructure to most disadvantaged nations.
NGO/NPO	Promote multi-country innovation and technology transfer. Focus funding on innovations that assure food and water security to most disadvantaged Support school and local government projects that address innovation that empowers people.
Private Sector	Privately owned and public stock companies and corporations have an obligation promote innovation that provides energy and water security, while promoting ecological balance and appropriate technologies that reduce CO_2 emissions and mitigate climate change.
Citizen Activist	Community organizers must see control of the innovation agenda as essential to addressing inequality and working toward sustainability. Activists must monitor the innovation agendas of all other sectors, particularly education and the private sector and make sure it is in line with the community's best interest.

Table 1 provides the various stakeholder groups recommended actions for promoting innovation. With stakeholders implementing the unified model of system of innovation on an Africa wide basic the development and deployment of appropriate technologies is possible on a continental scale.

Concluding Remark

In this chapter we have highlighted that the problem of African development has not been addressed as it should have been. The search for solutions within the existing fragmentations has exacerbated African underdevelopment rather than promote African development.

We looked at the theories of economic development and the internist self- critiques and found that even if there are improvements in thematic choice extension, when it comes to Africa the real dilemma is more related to finding a framework for understanding and explaining the evolutionary dynamics and processes of an integrated national African economy sustainable development innovation system.

We looked for possible inspiration from both systems of innovation and evolutionary perspective and suggested that there is much these theories can offer by reaching out to redress finally the core deficiency that Africa's development has suffered for centuries: the absence of an integrated and independent national development framework to address the problem of Africa's united transition from the current low-level of industrialization to a digitalization knowledge economy.

The task now is to re-think all the economic and innovation theories how to generate new knowledge in Africa. Beyond the issue of changing and reformulating the issues and themes for African development, theories that facilitate themes that connect peoples, activities, knowledge, projects across Africa are needed more than ever to try what Africa never tried to take the vision, mission, and tasks to unite. It is time for Africa to make the transition from the current underdevelopment state to enter into a fully highly developed integrated African sustainable development era freed from climate change, Covid-19 Pandemics and all the unending dependence on loans, debt and aid other risks.

A holistic innovation African system to integrate economics, space, politics, social, health, environment domains and spheres in Africa is much needed to create a sustainable and humane anchored smart development. At this time of the fourth industrial revolution we need to apply digital technology to create full trust, cooperation, and collaboration to realize the appropriate innovation where the political, social, economic and ecological spheres can align to validate the outcome that removes damage and failure and promotes success.

Africa needs an innovative theory of social capital now to unite us more than at any other time. If Africa unites, renascent Africa will emerge with the well-being of the people becoming the priority of priorities to make resource rich Africa in this planet serve the people in Africa and the crest of the world. It is time to engender an innovation of hope by uniting Africa and creating social technology, social innovation, and social entrepreneurship. A new culture and trust anchored collaboration with innovative, renascent, and united Africa is now more essential and necessary than at any other time with the risks of global climate change.

The contribution of the chapter is a unified and holistic African innovation system. It will challenge all the past and present problems Africa is facing as well as potential problems in the future. Stakeholder commitment to appropriate technology paired with this innovation system will empower communities across Africa. This pairing can bring development across Africa that will reduce poverty and increase well-being to hundreds of millions of people. At the same time these innovations can contribute to global sustainability and help mitigate climate change. This change in Africa can serve as a catalyst to other regions and convince stakeholders globally that this unified model of system innovation can be productive. The impacted communities can serve as a base to increase stakeholder sectors committed to an agenda of appropriate technology.

Bibliography

[1] D. Elleman, Helping People Help themselves: from the World Bank to an alternative philosophy of Development, Ann Arbor: University of Michigan Press, 2006.

[2] A. P. Thirlwall, Growth and Development with Special Reference to Developing Economics 8th edition, Palgrave Macmillan, 2006.

[3] A. Hirshman, "The rise and decline of Development Economics," in *Essays in Trespassing: Economics to Politics and Beyond*, Cambridge, Cambridge University Press, 1981.

[4] A. W. Lewis, The state of development Economics, London: Hobart, 1984.

[5] P. Krugman, "Towards a Counter-Revolution in Development Theory," in *Proceedings of the World Bank Annual Conference on Development Economics*, Washington DC, 1992.

[6] R. S. G. R. Perkins, Economics of Development 5th Edition, New Yorj: W W Norton & Company, 2001.

[7] K. Jomo, The Pioneers of Development Economics: great economists on Development, London & New York: Tulika Books & Zed books.

[8] P. Krugman, "What Economists can learn from Evolutionary Theorists?," European Association for Evolutionary Political Economy, 1996.

[9] J. Schumpeter, "Capitalism, Socialism and Democracy 5th edition," George Allan & Unwin, New York, 1976.

[10] G. &. L. Muchie, Putting Africa First: the Making of African Innovation Systems, Copenhagen: Aalborg University Press, 2003.

Epilogue

Bringing it All Together and Planning the way Forward

John Trimble and Mammo Muchie

> "According to figures from the World Bank, in 2021 the five hundred largest transnational corporations jointly controlled 52.8% of global GDP. They're obsessed with only one strategy: profit maximisation in the shortest time possible and at any price. ... The capitalist mode of production is undoubtedly the most energetic and creative ever known to humanity, but it's beyond the control of nation states or the unions. Corporations have a power that no king or pope could ever have dreamed of. What this ends up in is a cannibalistic world order. We're all fighting against hunger. We're fighting against the unlimited power of the multinational firms. We're fighting against global warming. [**Is capitalism responsible for climate change?**] Obviously. There's no public authority, in the name of the common good, that can enforce the normativity agreed by the participating states in the context of the Paris Agreement. The destruction of the planet is therefore the direct causal result of capitalism. Take the 2015 Paris Agreement. The aim was to limit global warming to 1.5°C compared to pre-industrial levels. Here we are in 2022, and instead of being reduced oil production has tripled for profit reasons." [1]

The above excerpt from a June 2022 interview with Jean Ziegler captures an assessment of the relationship between the capitalist free market system that dominates the world and the everyday problems faced by impoverished millions. The conquest mentality that has dominated the world over the past five centuries has resulted in the enslavement and genocide of millions. It has resulted in environmental degradation that is unimaginable. We cannot turn back the hands of time. Recognition of centuries of injustice is the first step in addressing the current inequalities. This manifesto hopes to motivate the various stakeholders to use their resources to commit to actively promoting appropriate technology. Technology policies that empower people must address the basic needs of the world's poorest communities. As Ziegler points out in his 2011 book *Betting on Famine Why the World Still Goes Hungry*

> "Rupture, resistance, the support of the world's peoples in opposing the established order are indispensable, at every level – globally and locally, in theory and in practice, here and everywhere. There must be concrete, voluntary acts, like those in which the farmers' unions in Ross Bethio, Benin, or in the high sierra of the Yucatan in Guatemala, or in the rice fields of Las Pavas, Columbia are engaged." [2, p. 247]

The tenth chapter of this manifesto speaks to the importance of human innovation in developing technology that empowers people. An innovation strategy focusing on Africa is presented as a guide for all stakeholders to embrace change. Innovation was presented in chapter one as a driving force in scientific and technological revolutions. Historically this innovation has been motivated by the conquest mentality explained in the first chapter. Currently innovations are funded primarily by multinational corporations. In the USA a major funding source for innovation and research is the department of defence and associated defence industries. It is the responsibility of the various stakeholders to take a humanist approach departing from the conquest mentality. In directing innovation in technology, an understanding is needed that the least fortunate and poorer communities must be given priority. Advocating humanism is clearly an issue of ethics.

A full chapter of this manifesto is devoted to the ethics of appropriate technology. It was written by a team of scholars from several universities. Three of the authors are founding members of INAT.

They present a historical overview of ethical systems, citing the contributions of African, Asian and European philosophers. The call for an ethical model of conduct when addressing technology development, is grounded in the work of Professor Godfrey Nzamujo. Nzamujo has implemented a network of rural ecovillages that promote autonomy and authenticity. These villages are examples of human-centred holistic collectives. Emphasis is on sustainability, training of local communities and maintenance of a circular economy. These ecovillages have also been established where land is available in urban settings. It is estimated that by 2050 over 70% of the world's population will live in cities. With this in mind, the proposed ethics model also addresses appropriate technology in dense urban settings. This model is a framework that future research and activists can build on. A concern for ethical practice and humanist driven decision making is expressed across the other chapters of the manifesto.

Industry is most closely associated with manufacturing. Advances in manufacturing have been critical in each of the four industrial revolutions. Machines have increased the productivity of manufacturing over the centuries and have made the supply chain from raw materials to end consumer products more complex. Labour remains the component responsible for all profits from manufacturing. The financial benefits from manufacturing, like other industries, go to the owners of the production environment not the workers or their communities. The chapter on manufacturing is a conversation between Professors Trimble and Mpofu. Transportation and the manufacture of transport vehicles receives special attention. This reflects Professor Mpofu role in heading the Gibela Research Chair (responsible for commuter rail research) and South African Research Chair (SARChi) in Future Transport Manufacturing Technologies. The chapter covers alternatives from solar powered cars to electrical vehicles. The potential of each stakeholder group influencing manufacturing is considered. The chapter ends by raising the paradox of electric cars being green technology. With coal-fired electric power generation in South Africa currently more greenhouse gas (GHG) is generated in the production and operation of an electric car than a comparable petrol car. Countries must plan advances in manufacturing and transportation with the well-being of its citizens in mind. Technology choices must be made that consider the health of people and the planet.

The opening chapter points out that mining is historically linked to the earliest conquest efforts that led to the enslavement and genocide of indigenous peoples, and extensive ecological degradation of mother earth. The chapter on mining provides personal insight on mining experiences in Guyana and highlights centuries long impact of mineral extraction in South Africa. South Africa's South Deep gold mine is the largest gold mine in the world with reserves of over 32 million ounces [3]. "The top mining companies saw their net profits increase by 15% in 2020 compared to the year before, and in 2021 profits more than doubled. ... The power imbalance between mining companies and local communities [increased] during the pandemic across several Latin American countries: Mexico, Honduras, Panama, Colombia, Ecuador, Peru, Brazil, Chile and Argentina" One example is when the Lagoa do Pirocáua dam failed in March 2021, the mining operation contaminated the water supply the Maranhão village in Brazil [4]. Across the global south, neo-liberal policies provide increased profits to multinational mining operations while the local communities suffer. Communities and activists must take a lead in forcing all stakeholder groups to address this injustice. Extensions to this manifesto must address the ongoing mining abuses and the potential to use technology to address this global inconsistency.

Farming and agri-industry are fundamental to ending hunger and starvation. Food security is a basic human right that can lead to reduction in inequality. Healthy people are more capable of addressing injustice and seeking empowerment. As the chapter of AT and agriculture states "In order to realize a sustainable future, we must look back, learn the lessons of the past and act accordingly". The chapter examines the role of agriculture in developed countries like the USA as well as parts of the global south, including South Africa, Haiti, Kenya and Ethiopia. The author points out "Appropriate agriculture strategies begin with healing the land and restoring biodiversity. It returns sovereignty to communities from the beginning to the end of the food chain, cutting their economic chains to debt, dependence and discrimination". An emphasis is on regenerative agriculture and food sovereignty. In 2007 the Nyeleni Declaration for Food Sovereignty was signed. The six principles of this declaration are presented in the chapter as foundation for continued work toward food sovereignty. The chapter ends with presenting the roles and responsibilities of each stakeholder group in the global drive to food sovereignty. Linking the work of many organizations across the global south is a critical ongoing challenge.

The service sector is the fastest employment sector. It represents a wide range of employment, includes human centred vocations to many related to conquest mentality, like military and police. The postal service was one of the first communication services connected on an international level. The extensive infrastructure accumulated over the years by postal systems in developed countries represented a commitment to community engagement. In recent years, with competing communication systems and rising private sector competition in parcel delivery, countries have shifted away from commitment to state-owned postal services. This is consistent with a neo-liberal agenda promoting privatization across a number of service industries, most notably energy generation and distribution and water and sanitation. The chapter on services highlights the dangers when service delivery is not accountable to the communities they engage. Service workers are part of these communities. The dominance of the service sector in developed countries places particular significance on service sector workers role not only in improving service delivery but also in addressing broader issues of wealth and income inequality. Developing countries have a smaller percentage of service sector workers compared to developed countries. Industrialisation is improving service delivery and increasing the proportion of service workers. The chapter covers seven basic needs service sectors. A summary of recommended actions is presented for each of the six stakeholder categories. It is critical that we heighten the research examining the dynamics and interrelated nature of these key service categories. How they may grow, decline and/or be transformed is linked to how technology is developed and deployed to impact each service category's relationship to communities.

This manifesto focuses on one particular service sector – water and sanitation. Safe water and sanitation is fundamental to a people's health. Water is essential to agriculture. Water is used in many manufacturing industries, mining and food processing, as well as in the generation of energy. The abuse of water resources has caused environmental damage, created unhealthy communities and destroyed many natural water sources. The chapter presents examples of appropriate technology in water harvesting, water purification and wastewater treatment. It captures the significance of planning the infusion of appropriate technology in delivering this key service. Other service sectors need to be examined – energy, communications, transportation, education, healthcare. In future releases of the AT Manifesto this expansion is a must.

History of technology development must be taught as it happened, in the context of conquest. The current imbalance in wealth, income and power is linked to this legacy. The first chapter of this manifesto provides a summary of conquest technology from the first arms factories of the 17th century; the expansion of shipbuilding to the shift to air warfare. "In 2021 world military expenditure surpassed the two trillion US dollar mark for the first time, reaching $2113 billion" [5]. The USA spent 38% of this. The military industrial complex and the terror of war is today's face of this conquest mentality. This reflects the highest form of inappropriate technology. In 2022, the leading world powers committed billions of dollars more to this military war machinery. The call is to shift away from inappropriate technology to appropriate technology. The voice of the oppress rings clear in this chapter that this shift is a necessary act of justice. The second chapter in the manifesto applied the scientific methodologies of dialectical materialism and system dynamics to further analysis technology development and deployment. Stock and flow diagrams provide conceptual models that convey the dynamic relations with appropriate and inappropriate technologies. Tools such as Stella and Vensim can be used to expand and quantify these models and present different potential scenarios. As this chapter concludes "Historical materialism identified a clear division in the interest of the owners of technology and the extensive populations engaged in developing and utilizing the wide range of technologies active in society. The identification of damages caused by extensive use of technologies should be a concern of all stakeholders". The six stakeholder groups: educational institutions; government structures; international agencies; non-governmental organizations, non-profit organizations and community groups; the private sector; and engaged citizens are presented in Chapter 3. The reasoning behind this classification is presented, as well as the contrasting responsibilities and roles of each stakeholder group. The chapter concludes by pointing out that the 'Citizen activists' is the most critical stakeholder and must serve as a catalyst to hold other stakeholders accountable.

The conquest mentality runs deepest in the minds and actions of some of the wealthiest men in the world. Billionaires feel justified in using wealth for personal gain because the conquest of wealth is glorified. The dominant ideology promotes that the power that comes from the control over resources is justified. The history of conquest shows its patriarchal nature. The process of conquest was initiated by European nations in the fifteen century and brought slavery, the slave trade, colonialism and settler colonialism. The results today are racism, white supremacy and white privilege. This nexus of white privilege, patriarchy and conquest mentality was very glaring in 2021 with three white men raced to conquer space. Elon Musk, born in apartheid South Africa, Jeff Bezos born in USA and Richard Branson born in the UK competed to be the first to take civilians into space. Billions of dollars were spent by their three companies to conquer space: Elon Musk's SpaceX, Jeff Bezos' Blue Origin and Richard Branson's Virgin Galactic. While billions of people don't have enough food, clean drinking water or healthcare, millions of people followed in glee these three efforts to conquer outer space. Why is this, and how can we redirect people's thinking and the world's resources?

The call to advance technology to empower people and increase the well-being of all of humankind, while striving for global ecological balance is the responsibility of all humankind, all stakeholder groups.

The way Forward is to Embrace the call for Appropriate Technology

As indicated in the introduction, the roots of this manifesto go back to the Appropriate Technology Declaration adopted at a National conference on Appropriate Technology in 2010. As an outgrowth of the AT declaration, this AT manifesto seeks to provide details on how our society has arrived at the current juncture of technology inequality.

Technology is a critical component of the forces of production. Technology combined with the people's labor and skills are primarily responsible for the production processes and services that advance society. It is important to promote technologies and production techniques that give workers and their communities more control over all production and distribution activities. This increase in control over the means of production is important in heightening class struggle.

Class struggle is needed to eliminate the inequalities inherent in capitalism. Working people and their communities must call on all stakeholders to embrace appropriate technology. By doing this an important shift in the forces of production takes place. In addressing technology changes priority is given to meeting the basic needs of the most disadvantaged communities, assuring human rights of all global citizens, and focusing on environmental integrity. At the same time, this stakeholder engagement leads to changes in relationships of production. This class struggle will lead to the needed qualitative leap in production relations, where control over production and distribution shifts from the hands of a few to those of the majority.

We must view this manifesto as a link, and extension of the AT declaration. At the same time it is living document. We must learn from its shortcoming. The AT manifesto is a work in progress and the responsibility of all stakeholders to see that it is clarified and expanded.

Bibliography

[1] J. Ziegler, "SWI SwissInfo.ch," 29 June 2022. [Online]. Available: https://www.swissinfo.ch/eng/jean-ziegler---the-thorn-in-the-side-of-swiss-capitalists/47706356. [Accessed 14 September 2022].

[2] J. Ziegler, Betting on Famine Why the World Still Goes Hungry, New York: The New Press, 2013.

[3] "Mining Technology," 21 July 2022. [Online]. Available: https://www.mining-technology.com/analysis/feature-top-ten-biggest-gold-mines-south-africa/. [Accessed 19 September 2022].

[4] I. V. Halm, "Mine," September 2022. [Online]. Available: https://mine.nridigital.com/mine_sep22/mining_profits_communities. [Accessed 20 September 2022].

[5] D. L. DaSilva, N. Tian, L. Déraud-Sudreau, A. Marksteiner and X. Liang, "SIPRI Fact Sheet," April 2022. [Online]. Available: https://www.sipri.org/sites/default/files/2022-04/fs_2204_milex_2021_0.pdf. [Accessed 21 September 2022].